THE ART OF WOODWORKING

SHOP-MADE JIGS AND FIXTURES

THE ART OF WOODWORKING

SHOP-MADE JIGS AND FIXTURES

TIME-LIFE BOOKS
ALEXANDRIA, VIRGINIA

ST. REMY PRESS
MONTREAL • NEW YORK

THE ART OF WOODWORKING was produced by
ST. REMY PRESS

PUBLISHER	Kenneth Winchester
PRESIDENT	Pierre Léveillé
Series Editor	Pierre Home-Douglas
Series Art Director	Francine Lemieux
Senior Editors	Marc Cassini (Text) Heather Mills (Research)
Art Directors	Normand Boudreault, Luc Germain, Solange Laberge
Designers	Lina Desrochers, Hélène Dion, Jean-Guy Doiron, Michel Giguère
Research Editor	Jim McRae
Picture Editor	Christopher Jackson
Writers	Andrew Jones, Rob Lutes
Research Assistant	Bryan Quinn
Contributing Illustrators	Gilles Beauchemin, Roland Bergerat, Michel Blais, Jean-Pierre Bourgeois, Ronald Durepos, Robert Paquet, James Thérien
Administrator	Natalie Watanabe
Production Manager	Michelle Turbide
System Coordinator	Jean-Luc Roy
Photographers	Robert Chartier, Christian Levesque
Administrative Assistant	Dominique Gagné
Proofreader	Judith Yelon
Indexer	Christine M. Jacobs

Time-Life Books is a division of Time Life Inc.,
a wholly owned subsidiary of
THE TIME INC. BOOK COMPANY

TIME LIFE INC.

President and CEO	John M. Fahey
Editor-in-chief	John L. Papanek

TIME-LIFE BOOKS

President	John D. Hall
Vice-President, Director of Marketing	Nancy K. Jones
Executive Editor	Roberta Conlan
Executive Art Director	Ellen Robling
Consulting Editor	John R. Sullivan
Production Manager	Marlene Zack

THE CONSULTANTS

Jon Arno is a consultant, cabinetmaker, and freelance writer who lives in Troy, Michigan. He also conducts seminars on wood identification and early American furniture design.

Kam Ghaffari is a freelance writer and editor. He has his own business in Rhode Island designing and building one-of-a-kind and limited production furniture. Kam's background also includes working professionally in furniture reproduction and fine carpentry and studying with furniture patriarchs Wendell Castle of the U.S. and England's Fred Baier.

Giles Miller-Mead taught advanced cabinetmaking at Montreal technical schools for more than ten years. A native of New Zealand, he has worked as a restorer of antique furniture.

Joseph Truini is Senior Editor of *Home Mechanix* magazine. A former Shop and Tools Editor of *Popular Mechanics,* he has worked as a cabinetmaker, home improvement contractor, and carpenter.

Shop-made jigs and fixtures
p. cm.—(The Art of Woodworking)
Includes index.
ISBN 0-8094-9508-2 (trade)
1. Jigs and fixtures.
I. Title: Shop-made jigs and fixtures.
II. Series
TJ1187.S54 1993
684'.083—dc20 93-34103
CIP

For information about any Time-Life book, please call 1-800-621-7026, or write:
Reader Information
Time-Life Customer Service
P.O. Box C-32068
Richmond, Virginia
23261-2068

CONTENTS

R. J. De Cristoforo on

DESIGNING JIGS

I look at a new tool as a beginning. Once it is taken from its box or crate I read the owner's manual to learn what the manufacturer suggests the tool can do. Then I stand back and think, "There must be more to it than this."

Inevitably, by some strange thought process I can't explain, there appears a mental picture of a jig, sometimes simple enough to test immediately, other times elaborate enough to require a session at what at one time was a drawing board; now, I design on my computer. The new jig might enable the tool to do something its designer never envisioned, or it might increase accuracy with minimum fuss, or it could add a safety factor to a routine operation. In any case, it has to be custom-made since it is rarely available commercially.

I've designed dozens of jigs for power and hand tools. Still, for me it's not an obsession: Practicality is essential and shop testing must prove the jig's worth. Some folks think that jigs are only for amateurs. If so, there are many professionals working in amateurish ways!

Jigs are meant to be used. Those that I design are not made for the sake of a magazine story or a book and then stored or discarded. In a sense, I conceive a project that helps me exploit a machine, or assists me in working more accurately and safely, and then I share it with other woodworkers. If I've proven that a jig will be useful to just one reader then it has value for me.

I'm fond of the master jigs that I have made for the drill press and band saw, and especially the unit for the table saw shown in the photo at left. Its basic component is a generous sliding table with removable inserts so it can function with a dadoing tool as well as a saw blade. Its attachments include adjustable guides for accurate crosscutting and mitering, and a mountable unit—a jig in itself—that allows cuts like tenons and slots in the end of narrow stock. The device includes a number of essential but usually separate jigs and adds the advantages of a sliding table to each of them.

There's no doubt that jigs can help any woodworker, but they must be made carefully. These are situations where it pays to take 10 minutes to do a five-minute job. Consider that the jig will be a lifetime tool and you'll agree that making it right is the only way to go.

R. J. De Cristoforo, author of numerous books on woodworking and other subjects, lives in Los Altos Hills, California.

Ted Fuller and his
ROUTER JIG

Years ago, I worked in an auto body shop, where we hand-formed body panels. Often we needed two matching panels—one for each side—but we never produced exact mirror images. Subtle differences were easily excused: "You can't see both sides at the same time," it was said.

In cabinetmaking, however, matching pieces must be exact duplicates. You usually can see them at the same time. Although some masters can accomplish this freehand, most of us must rely on carefully constructed jigs.

I was lured to woodworking in grade school when the shop teacher put me to work on the props for a Christmas play. I learned early that the time spent on the jig or template meant time saved and consistency gained.

There are plenty of jigs and fixtures on the market but, like baseball gloves, only your own has just the right fit. When you make a jig yourself, it is designed for a specific application and sized to match your project. Best of all, you don't have to change the project to fit the store-bought fixture. It's also less expensive.

Making arched-top raised panel doors is one example. A manufactured jig, and pre-cut templates that enable you to do the job cost several hundred dollars. For the custom piece I am working on in the photograph I built a simple template out of plywood to create the contour for the arched top rail. This particular jig is adjustable for two widths of doors. Wider or narrower doors will require another jig and a re-draft of the curve.

With the multitude of top-bearing router bits on the market, exact contours are quick and easy to duplicate. Simply rough out the piece to shape, clamp on the template, and rout to the finished shape. Another benefit of this form of duplication is that it does not leave the tool marks that a bandsaw would and therefore reduces sanding considerably. The next step is to run the pieces through a matched set of rail and stile cutters to rout the profile on the sticks and make the cope cuts. It's a good idea to mark the dimensions, bit selection, and set-up information on the jig so recalculation is not required.

I think one of the most intriguing things about woodworking is that there is always some rigging that will make the work easier, faster, and better. The only real limitation is your own imagination.

Ted Fuller is the product manager at Delta International Machinery/Porter-Cable (Canada) in Guelph, Ontario. He is currently working in new product development and marketing for woodworking tools and equipment. He is also a keen woodworker.

Bruce Beeken and Jeff Parsons discuss

PLANNING JIGS

Every woodworker uses jigs regularly. Marking gauges, combination squares, the rip fence on a table saw, and router bits with ball-bearing pilots are all jigs that are taken for granted. And who hasn't, at one time or another, made a simple thingamajig on the spur of the moment to help get a certain job done?

In our shop, we design most of the furniture we make. In developing a new piece, we consider the esthetic and the building process at the same time. Our chairs, for example, have parts that don't come straight from machine tables because we like them to have a certain stance to support the person sitting on them just so.

Some of our more complex jigs are used in chair making. These sorts of jigs are planned from the outset, tailoring the process to the design. We develop them on a full-scale drawing as we work out our concept of the furniture piece. A clear understanding of the steps and their sequence not only makes the whole job less intimidating but often suggests ways to simplify the procedure and refine the piece itself.

Sometimes a jig is as simple as a wedge to jack up a part at the proper angle. We have also found that jigs can serve more than one purpose, traveling with a part from machine to machine. The home-made device shown in the photo, for example, is used for both forming chair legs on the shaper and cutting mortises in them on the mortiser. We start by band sawing blanks to approximate size. They are then fastened to the jig—in pairs, since the jig has two edges. We shape the inside faces of the legs. Then, by changing shaper knives, shifting the dowel pegs in the jig, and repositioning the legs, we can use the same jig to shape the feet. Once all the parts have been formed, we return the pegs to the face-shaping position and bolt the jig to our mortiser's table. The jig then holds the legs in the proper position as the mortises are cut.

There are times that, with a little extra effort, a jig can be made to serve a general purpose: for instance, a hinged taper jig for the table saw or thickness planer, or a router boom for cutting arcs.

The use of jigs is inseparable from our understanding of how to make the furniture we design. Even if a piece is to be made just once, it is likely that we will develop and use a jig somewhere along the line. When we are producing a batch of several hundred chairs, jigs are critical in almost every move we make. Whether simple or complex, they serve as the link between drawing and tool, ensuring consistent, precise results.

Bruce Beeken and Jeff Parsons are graduates of Boston University's Program in Artisanry. They build fine furniture at their shop at Shelburne Farms, Shelburne, Vermont.

ROUTING AND SHAPING JIGS

Since its invention early in the 20th Century, the router has become one of the most popular portable power tools—and with good reason. Few tools can match its speed, accuracy, and versatility for shaping wood or cutting joints. But jigs are almost a necessity; although the router can be used freehand, most cuts require a guide—particularly repeat cuts.

The jigs featured in this chapter provide various ways of obtaining quick and precise results from your router. Some, like the dadoing jigs shown beginning on page 16, reduce the setup time for simple procedures. Others, like the lap joint jig on page 27, allow the tool to produce multiple copies of the same joint in a few minutes. A relatively new woodworking development, vacuum-powered accessories (*page 34*) eliminate the need for conventional clamps when routing patterns using a template. The vacuum pump is also useful for securing featherboards to a router table. All of these jigs are easy and inexpensive to build.

The router's larger cousin, the shaper, can perform many operations better than the smaller tool, but it is generally regarded as one of the most dangerous tools in the typical woodshop. A shop-made featherboard *(page 32)* and guards *(page 33)* will make it a safer tool.

Aided by a simple jig, a table-mounted router cuts a perfect box joint.

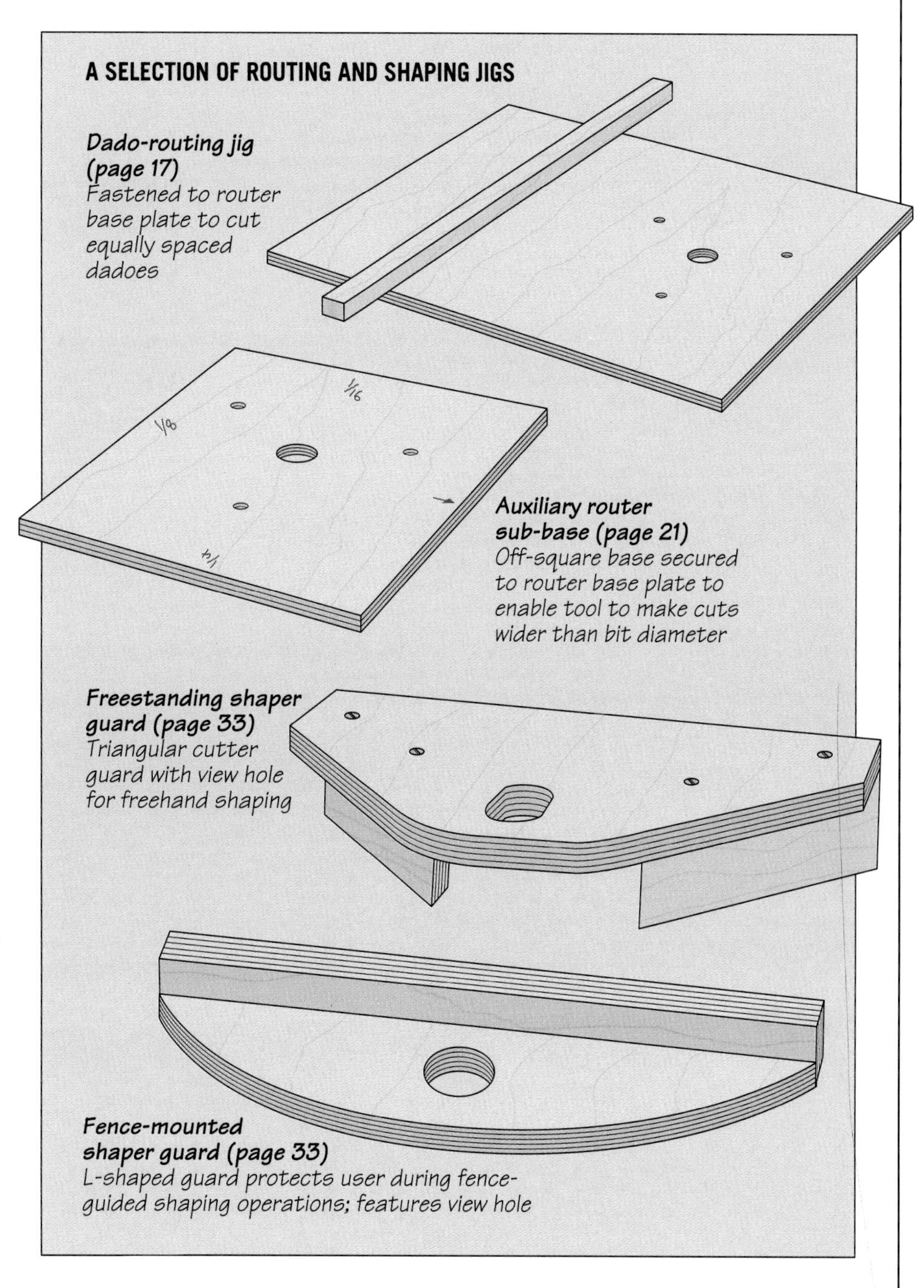

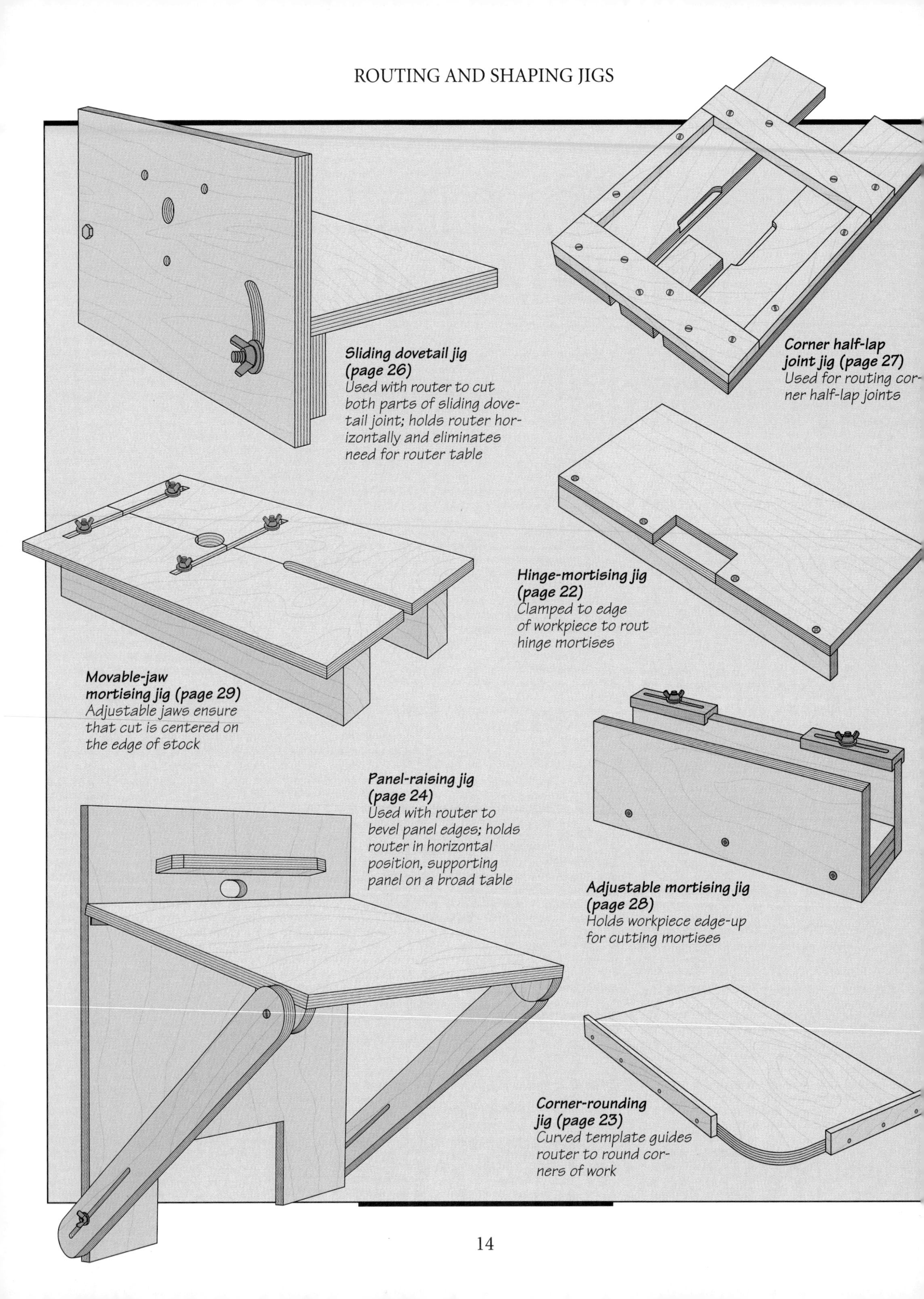

Sliding dovetail jig (page 26)
Used with router to cut both parts of sliding dovetail joint; holds router horizontally and eliminates need for router table

Corner half-lap joint jig (page 27)
Used for routing corner half-lap joints

Hinge-mortising jig (page 22)
Clamped to edge of workpiece to rout hinge mortises

Movable-jaw mortising jig (page 29)
Adjustable jaws ensure that cut is centered on the edge of stock

Panel-raising jig (page 24)
Used with router to bevel panel edges; holds router in horizontal position, supporting panel on a broad table

Adjustable mortising jig (page 28)
Holds workpiece edge-up for cutting mortises

Corner-rounding jig (page 23)
Curved template guides router to round corners of work

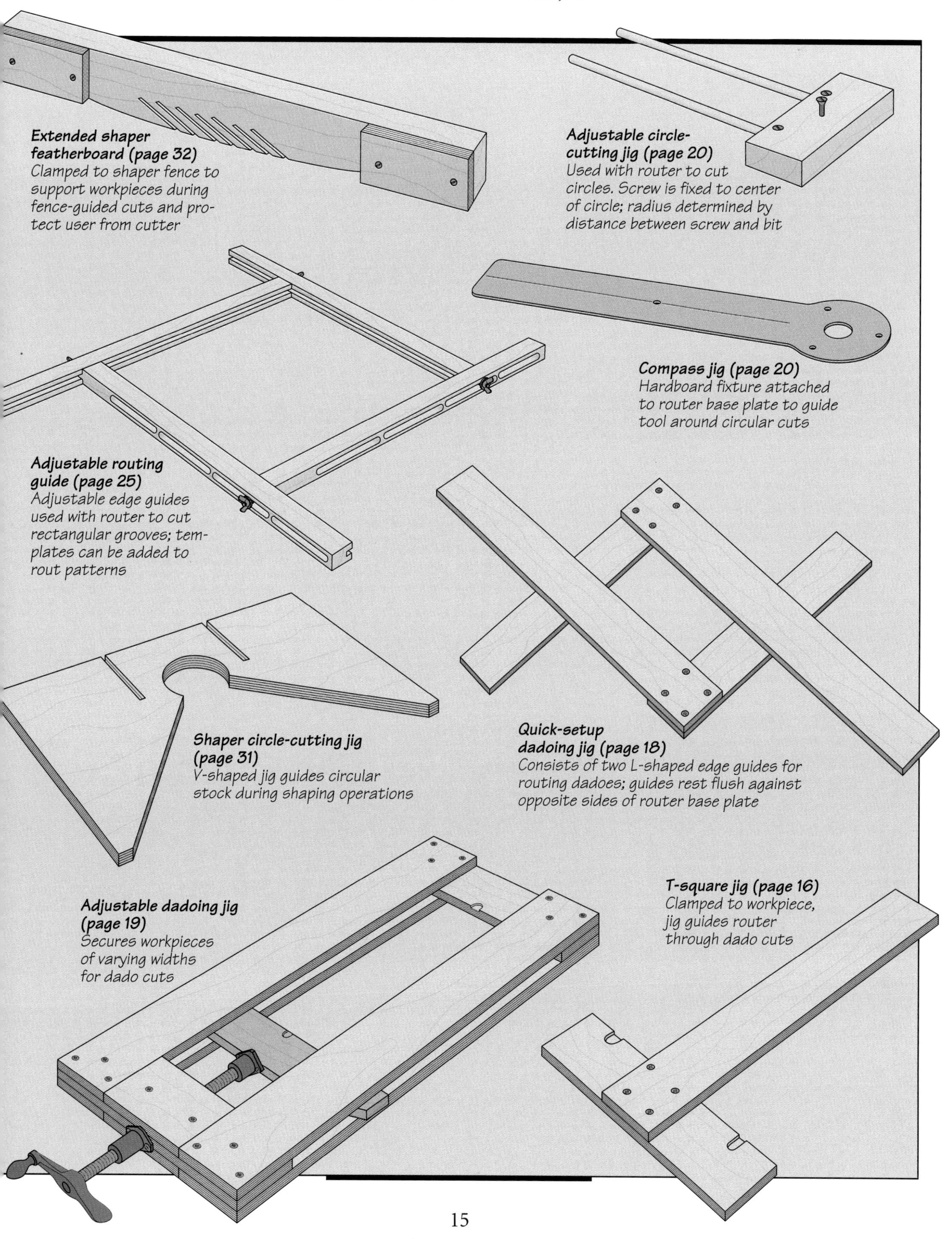
Extended shaper featherboard (page 32)
Clamped to shaper fence to support workpieces during fence-guided cuts and protect user from cutter
Adjustable circle-cutting jig (page 20)
Used with router to cut circles. Screw is fixed to center of circle; radius determined by distance between screw and bit
Compass jig (page 20)
Hardboard fixture attached to router base plate to guide tool around circular cuts
Adjustable routing guide (page 25)
Adjustable edge guides used with router to cut rectangular grooves; templates can be added to rout patterns
Shaper circle-cutting jig (page 31)
V-shaped jig guides circular stock during shaping operations
Quick-setup dadoing jig (page 18)
Consists of two L-shaped edge guides for routing dadoes; guides rest flush against opposite sides of router base plate
Adjustable dadoing jig (page 19)
Secures workpieces of varying widths for dado cuts
T-square jig (page 16)
Clamped to workpiece, jig guides router through dado cuts

A T-SQUARE JIG

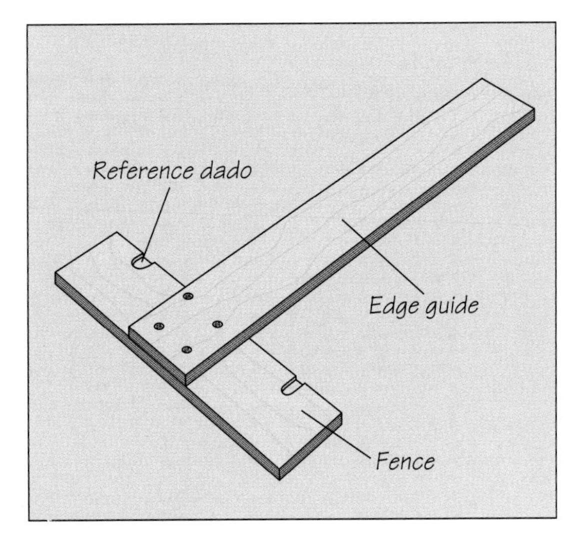

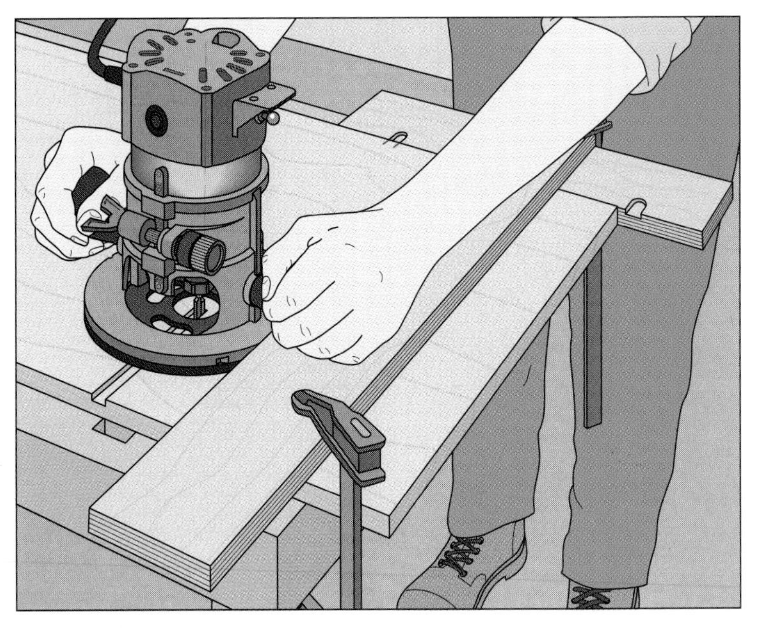

1 Building the jig

To rout dadoes that are straight and square to the edges of your stock, use a T-square jig like the one shown above. Make the jig from ¾-inch plywood, sizing the pieces to suit the stock you will be using and the diameter of your router base plate. The edge guide should be about 4 inches wide and longer than the workpiece's width; the fence, also about 4 inches wide, should extend on either side of the guide by about the width of the router base plate. Assemble the jig by attaching the fence to the edge guide with countersunk screws. Use a try square to make certain the two pieces are perpendicular to each other. Then clamp the jig to a work surface and, with the base plate against the edge guide, rout a short dado on each side of the fence with your two most commonly used bits—often ½ and ¾ inch. These dadoes in the fence will minimize tearout when the jig is used and help align the jig with the cut you wish to make.

2 Routing a dado

Clamp the jig to the workpiece, aligning the dado in the fence with the outline on the stock. When making the cut, press the router base plate firmly against the edge guide *(above)*. Continue the cut a short distance into the fence before stopping the router.

SHOP TIP

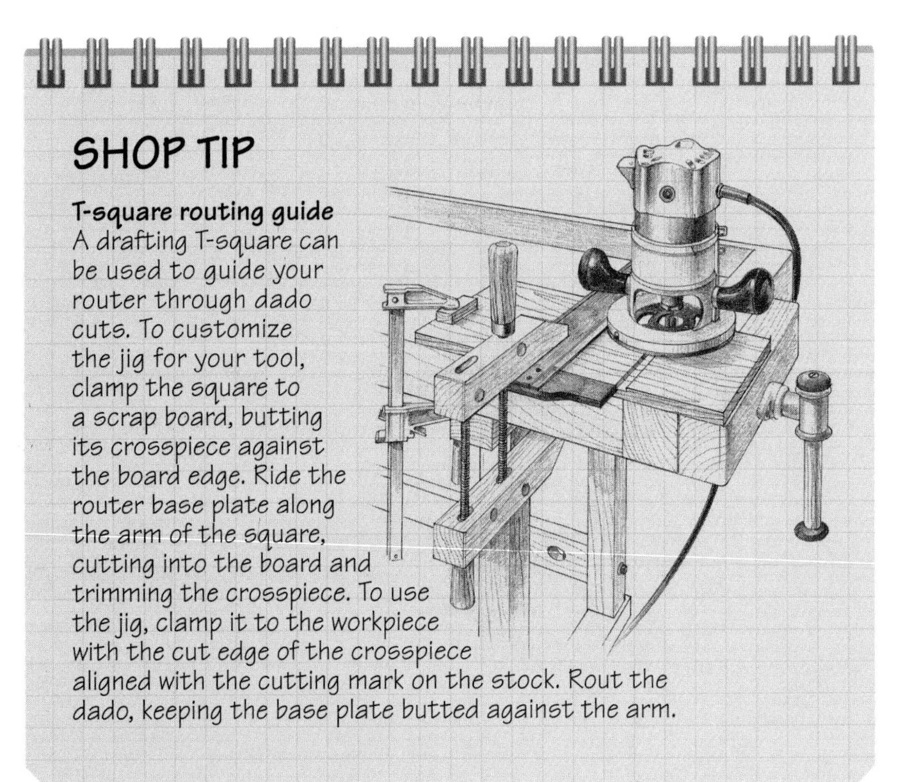

T-square routing guide

A drafting T-square can be used to guide your router through dado cuts. To customize the jig for your tool, clamp the square to a scrap board, butting its crosspiece against the board edge. Ride the router base plate along the arm of the square, cutting into the board and trimming the crosspiece. To use the jig, clamp it to the workpiece with the cut edge of the crosspiece aligned with the cutting mark on the stock. Rout the dado, keeping the base plate butted against the arm.

A JIG FOR EVENLY SPACED DADOES

1 Assembling the jig
The jig shown at right is ideal for cutting equally spaced dadoes with a router. Dimensions depend on the size of the workpiece and the spacing between the dadoes. Begin by cutting a piece of ¼-inch plywood for the base, making it a few inches wider than the diameter of your router's base plate and a few inches longer than the spacing between the dadoes. Set the base on a work surface and place your router near one end. Mark the screw holes in the router base plate on the base; also mark a spot directly below the tool's collet. Bore holes for the screws and cut a hole at the collet mark large enough for the router bit. Remove the sub-base from the tool, screw the jig base to the router base plate, and install a straight bit the same width as the dadoes you wish to rout. Next, cut a spacer to fit snugly in the dadoes, making it slightly longer than the width of the workpiece. Screw the spacer to the bottom of the jig, making the distance between it and the bit equal to the space you want between your dadoes.

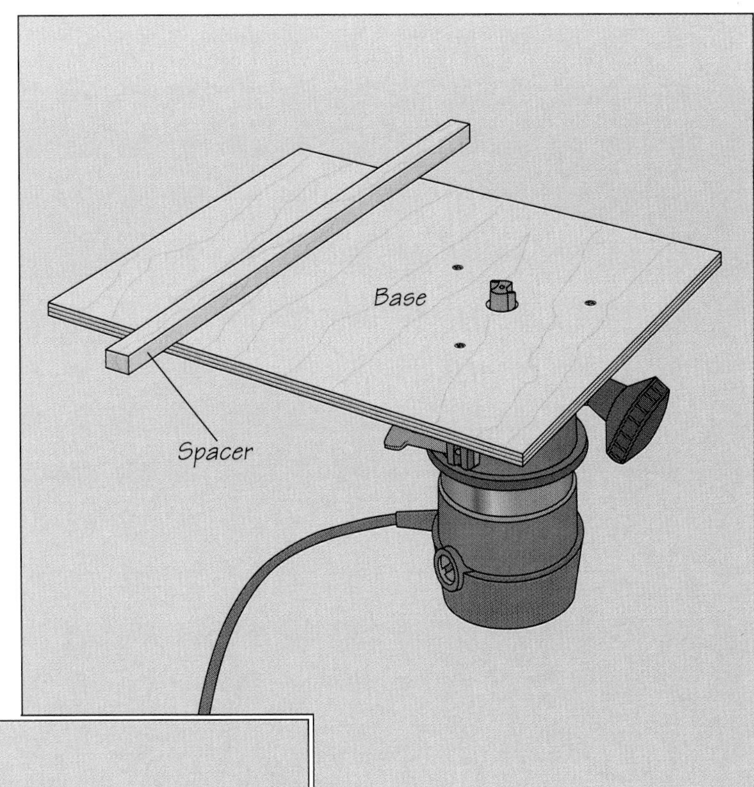

2 Cutting the dadoes
Clamp your stock to a work surface and set the jig on the workpiece with the spacer flush against one end and the router bit at one edge. Hold the router firmly and feed it across the surface to rout the first dado, keeping the spacer flush against the workpiece. Turn off the router and insert the spacer in the dado, repositioning the clamps as necessary. Rout the next dado, sliding the spacer in the first dado. Continue *(left)* until all the dadoes have been cut. (To vary the location of your first dado, rout it with a T-square guide like the one shown on page 16, rather than with the spacer jig.)

A QUICK-SETUP DADO JIG

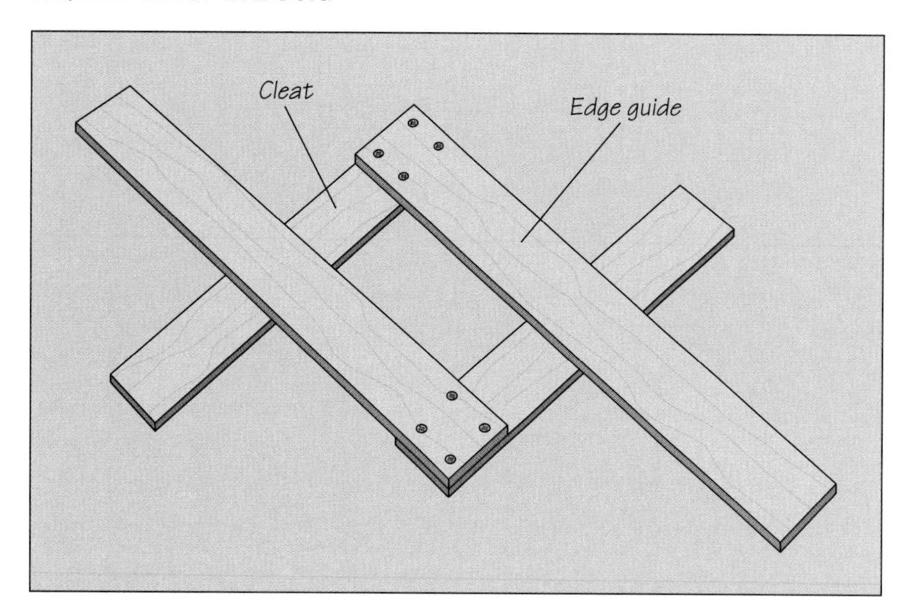

1 Assembling the jig
The jig shown at left makes it easy to rout dadoes with minimal tearout. The device consists of four strips of ¾-inch plywood attached to form two Ls. Rip all the pieces of the jig about 4 inches wide. Cut the edge guides a few inches longer than the cut you intend to make. The cleats should be long enough to overlap the adjacent edge guide by several inches when the jig is set up. Fasten the edge guides to the cleats, making certain the pieces are square; use four countersunk screws for each connection.

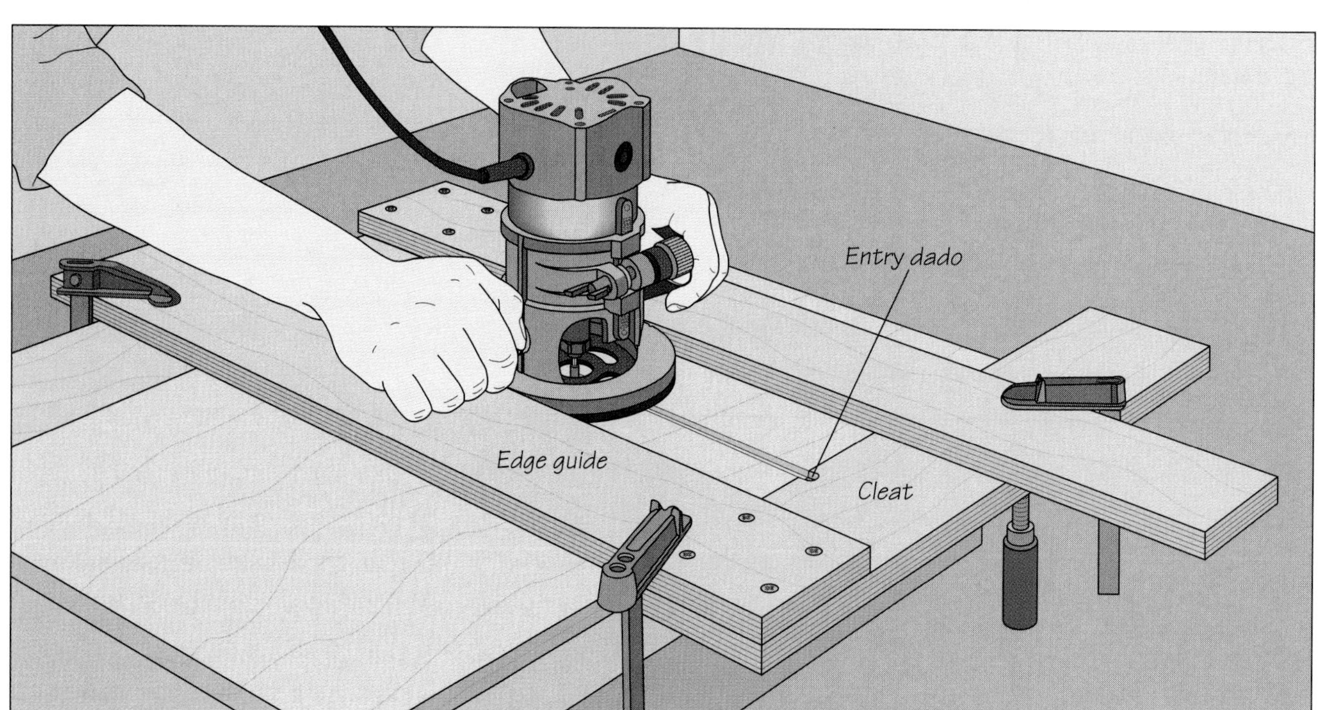

2 Routing a dado
Set up the jig by clamping the stock to a work surface and butting the cleats against the workpiece at the beginning and end of the cut. Then set your router between the edge guides, aligning the bit over the dado outline. Slide the guides together until they butt against each side of the router base plate. Secure the jig by clamping the Ls to each other and to the workpiece. Turn on the router and, with the tool between the edge guides, plunge the bit into the cleat at the start of the cut to form an entry dado. Guide the router across the workpiece *(above)*, extending the cut completely through the stock and into the second cleat. This will minimize tearout as the bit exits the workpiece. If you need to rout several dadoes of the same size, leave the jig clamped together and align the entry dado with the dado outline marked on the stock.

AN ADJUSTABLE DADO JIG

1 Building the jig
The jig shown at right is ideal for routing dadoes in wide panels. Size the pieces so the distance between the edge guides equals the diameter of your router's base plate. The guides should be long enough to accommodate the widest panel you plan to rout. Cut the four edge guides, the two ends, and spacers from ¾-inch plywood; make all the pieces 4 inches wide. Sandwich the end pieces between the guides and screw them together. At one end of the frame, attach spacers to the top and bottom of the end piece. Countersink all fasteners. Cut the clamping block from ¾-inch-thick stock; make it about 3 inches wide and longer than the end pieces. To install the press screw, bore a hole for the threads through the end piece with the spacers *(right, below)*. Remove the swivel head from the press screw and fasten it to the middle of the clamping block. Attach the threaded section to the swivel head and screw the collar to the end piece. Use the router to cut short reference dadoes in the other end piece and the clamping block.

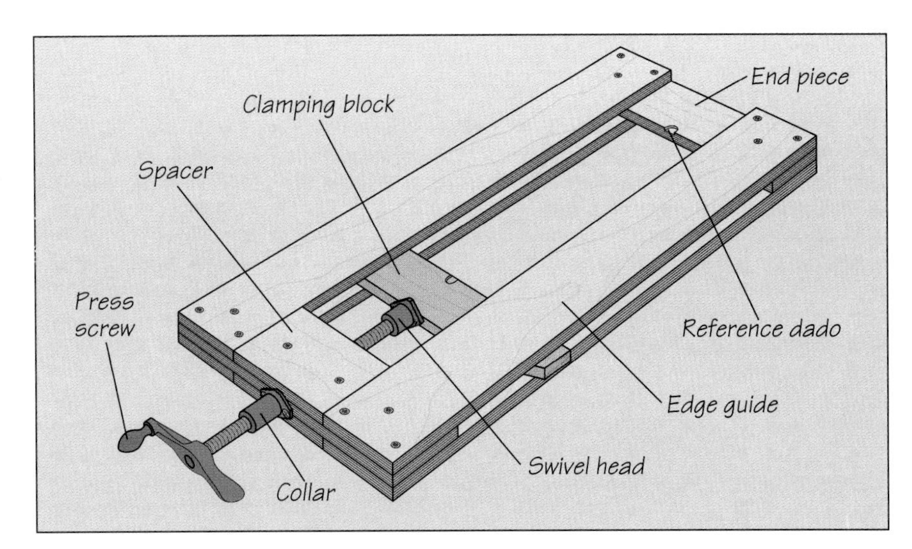

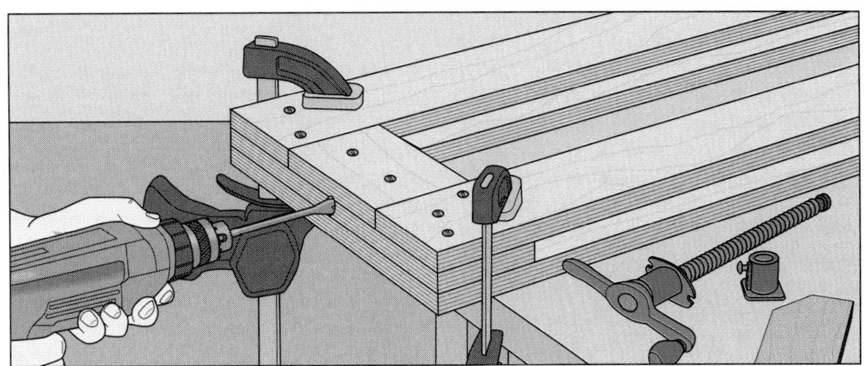

2 Cutting a dado
Slide the workpiece between the edge guides, aligning the marked outline with the reference dadoes. Secure the panel in position with the clamping block. Clamp the jig to a work surface. With the bit clear of the stock, turn on the router and start the cut at the reference dado in the end piece, making certain the router is between the edge guides. Feed the bit into the workpiece, keeping the base plate flat on the stock *(left)*. To minimize tearout, wait until the bit enters the reference dado in the clamping block before raising the router clear of the stock.

CIRCLE-CUTTING JIGS

Comprising two hardwood dowels and a center block, this adjustable jig allows your router to cut circles of virtually any diameter. The jig is assembled by slipping the dowels into the accessory holes in the router base plate, fixing the dowels to the block, and attaching the block to the center of the circle marked on the workpiece. With wood cleats holding the stock to a work surface, the router bit is aligned with the end of the circle's marked radius and the screws that clamp the dowels to the base plate are tightened. The circle can then be routed.

COMPASS JIG

Making the jig and routing a circle

To cut larger circles than most commercial guides allow, use the compass jig shown at right. Make the device from 1/4-inch hardboard, sizing it to suit your router. The router end of the jig should be circular and about the size of your tool's base plate. Make the arm at least 2 inches wide and longer than the radius of the circle you will be cutting. Bore a hole in the center of the rounded end for the router bit. To mount the jig on your router, remove the sub-base and center the bit over the clearance hole. Mark the screw holes on the jig, bore and countersink them, then screw the jig to the router. Draw a line down the center of the jig arm and mark the radius of the circle on it, measuring from the edge of the bit. Drill a hole at the center mark and screw the jig to the workpiece. Secure the stock to the work surface with cleats. Plunge the bit into the stock and rout the circle in a clockwise direction.

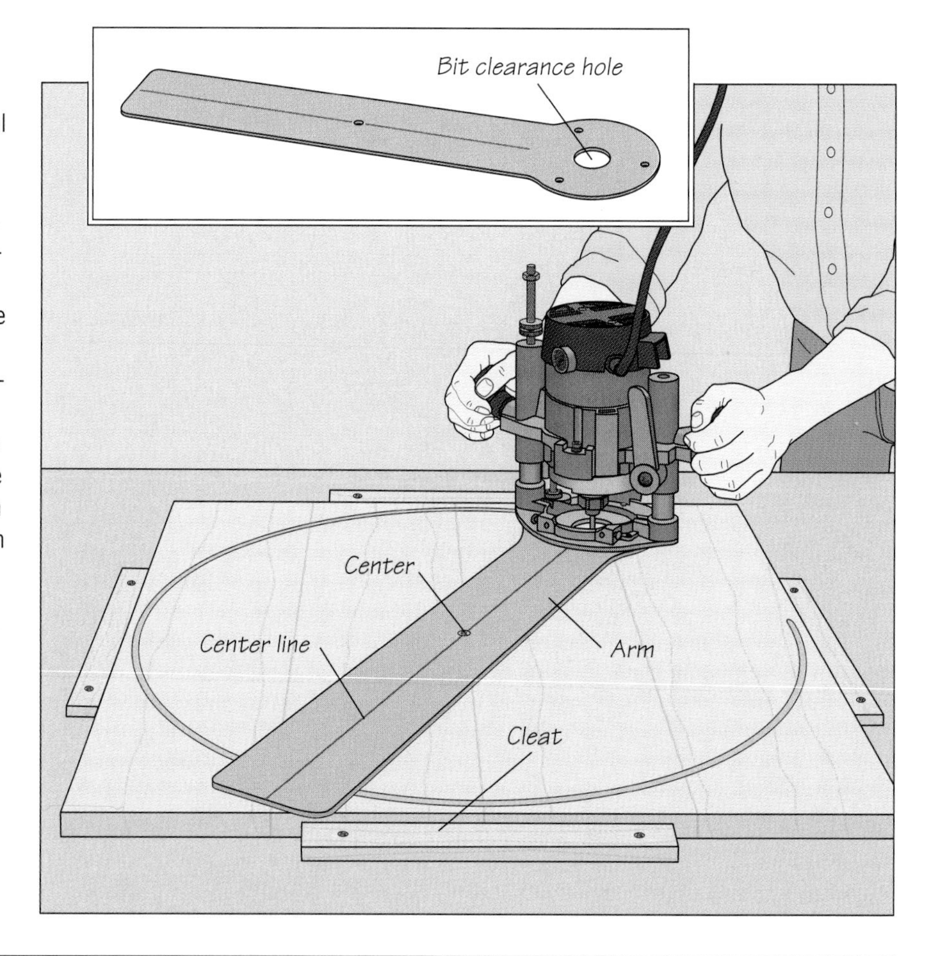

ROUTER JOINTING JIG

ROUTER JOINTING

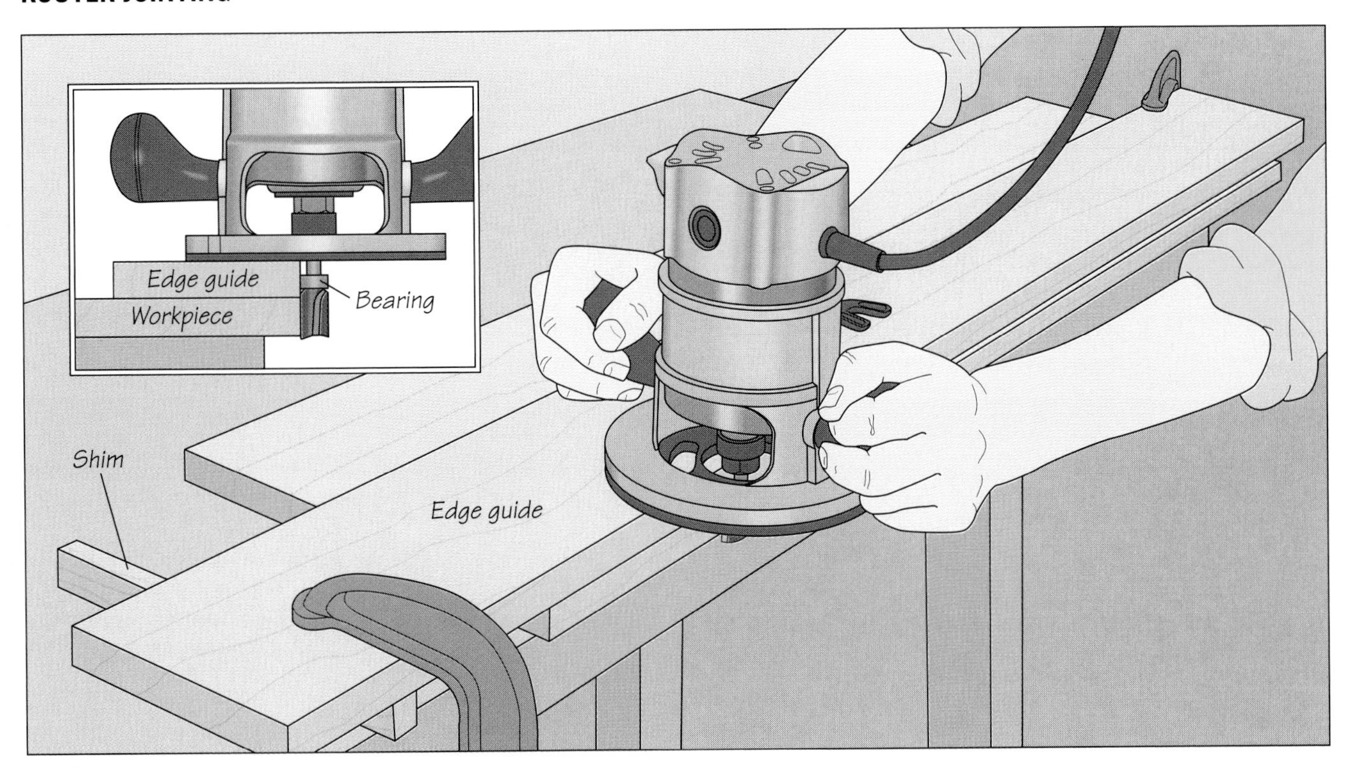

Jointing wide stock

To surface boards that are too cumbersome to move across the jointer, use a router along with a perfectly square edge guide. This technique works best using a top-piloted straight bit with a ½-inch shank. Position the edge guide atop the board to be jointed with the edge of the board protruding from the guide's edge by about 1/16 inch. Clamp both to the work surface. Make the edge guide longer than the workpiece to prevent the clamps from interfering with the router; place a shim under the clamp jaw to keep the guide from wobbling. With the router flat on the guide, adjust the bit height so it will cut the entire edge of the board *(inset)*. Feed the router from one end of the board to the other; the pilot will ride along the guide as the cutter trims the workpiece flush *(above)*.

SHOP TIP

An auxiliary sub-base for wide cuts

To make a cut that is wider than a particular router bit, you would normally make one pass, shift your edge guide and make a second pass. Or, you can use this auxiliary off-square sub-base. Cut a piece of ¼-inch plywood into an 8-inch square. Remove the router's sub-base and bore the screw holes and clearance hole for the bit through the auxiliary sub-base so the tool will be centered on the jig. Next, cut 1/16 inch of wood from one edge of the sub-base, ⅛ inch from an adjacent edge, and ¼ inch from a third edge. Mark the amounts you removed on each side. Screw the jig to the router and make a pass with the uncut end flush against the guide. Rotate the sub-base and make a second pass, widening the groove by 1/16, ⅛, or ¼ inch, depending on which side you use.

HINGE MORTISING JIG

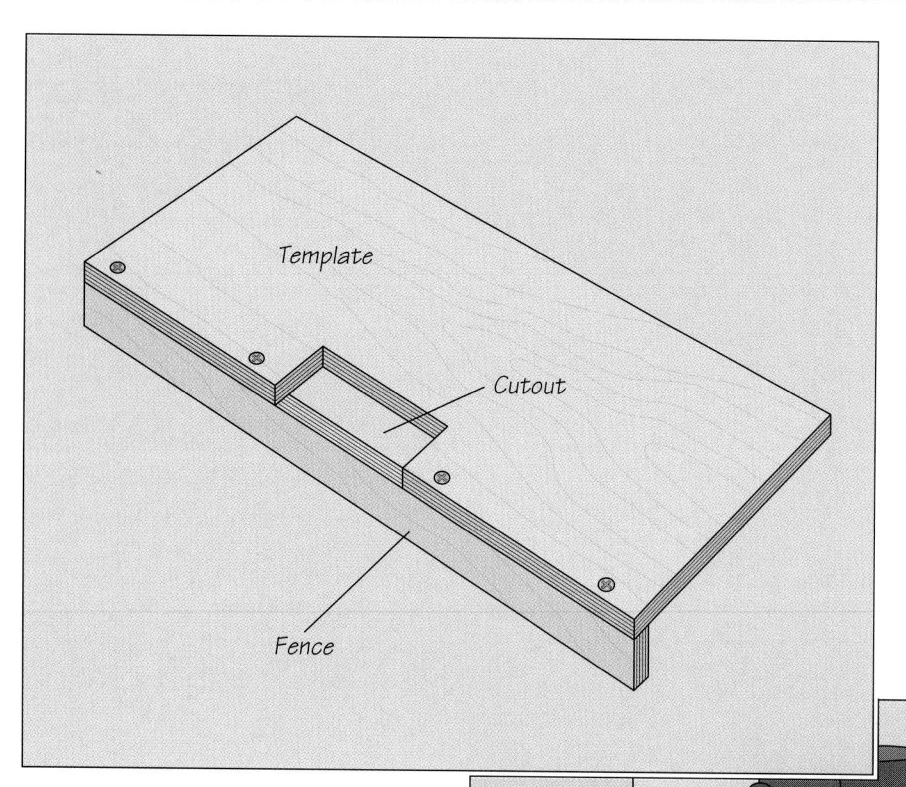

1 Building the jig
A jig like the one shown at left will allow your router to cut hinge mortises quickly and accurately. To make the cuts, you will need to equip your router with a straight bit and a template guide. Build the template from a piece of 3/4-inch plywood. Size it wide enough to support the router. Outline the hinge leaf on the template; remember to compensate for the template guide and the thickness of the fence, which is also made from 3/4-inch plywood. Cut out the template, then attach the fence with countersunk screws.

2 Routing hinges
Secure the workpiece edge-up in a vise. Mark the hinge outline on the stock and clamp the template in position, aligning the cutout with the outline on the edge and butting the fence against the inner face of the workpiece. Make the cut *(right)* by moving the router in small clockwise circles, then remove the jig and square the corners with a chisel.

CORNER-ROUNDING JIG

1 Constructing the jig
For curving the corners of a workpiece, you can use the simple corner-rounding jig shown at right. The jig consists of a plywood base and two lips that align the edges of the jig and the workpiece. Cut the base from 3/4-inch plywood. For most jobs, a base about 10 inches wide and 16 inches long will be adequate. Draw the curve you wish to rout on one corner of the base and cut it with a band saw or a saber saw; sand the edge smooth. Cut the lips from solid stock 1/2 inch thick and 1 1/2 inches wide, then nail or screw the pieces to the base, leaving about 3 to 4 inches between each lip and the rounded corner. The top edge of the lips should be flush with the top surface of the base.

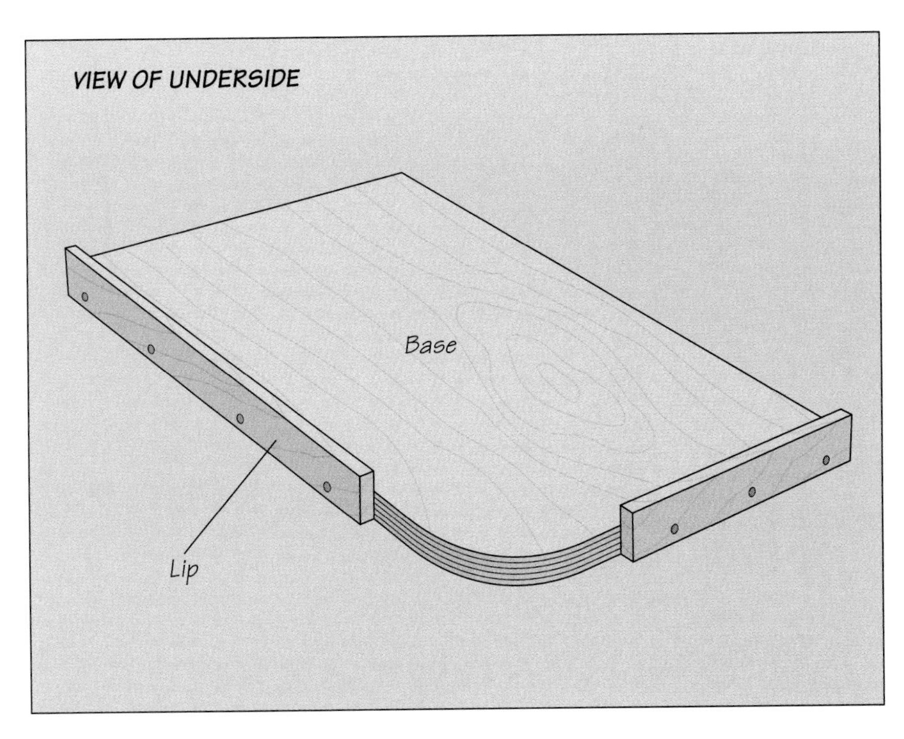

2 Rounding a corner
Set your stock on a work surface with the corner to be rounded extending off the table by several inches. Place the jig on top of the workpiece so the lips are butted against the edges of the stock. Use clamps to secure the two pieces to the work surface. To make the router cut easier, use a handsaw to cut away the bulk of the waste. Then, using a top-piloted flush-cutting bit in your router, start clear of the corner and ease the bit into the stock until the pilot contacts the edge. Pull the router around the corner, moving against bit rotation and pressing the pilot flush against the edge of the jig throughout the operation *(left).*

PANEL-RAISING JIG

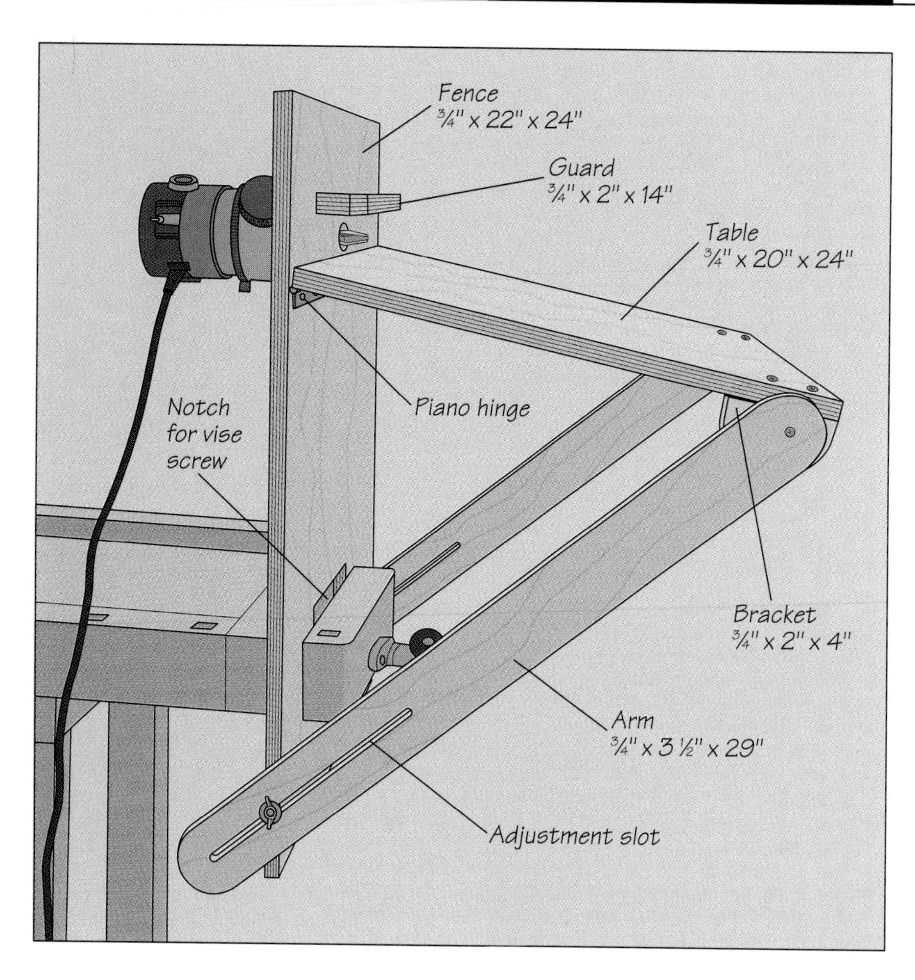

1 Building the jig
Featuring a fence and tilting table, the jig shown at left enables you to raise panels using a router without mounting the tool in a table. The jig is clamped in a bench vise. Cut all the pieces of the jig from ¾-inch plywood; the dimensions suggested in the illustration will work well with a typical workbench. Start assembling the jig by screwing the brackets to the underside of the table at one end, then cut adjustment slots through the arms. Secure the top ends of the arms to the brackets and the bottom ends to the fence using hanger bolts, washers, and wing nuts. Attach the table to the fence with a piano hinge positioned about 6 inches below the top of the fence. To prepare the fence for your router, bore a hole just above the table level to accommodate the largest ½-inch vertical panel-raising bit or straight bit you will be using. Screw the guard to the fence above the hole. Finally, cut a notch in the bottom end of the fence to clear the workbench's vise screw. Install the bit in the router and screw the tool to the jig fence so the bit protrudes from the hole.

2 Raising a panel
Secure the jig fence in the vise with the table at a comfortable height. Adjust the bit for a shallow cut, turn on the router, and make a test cut in a scrap piece. To adjust the bevel angle, turn off the tool, loosen the wing nuts securing the arms to the fence, and tilt the table up or down. Raise the ends of the panel before routing the sides; this will reduce tearout. Feed the panel across the table face-up, keeping your fingers well clear of the bit. Test-fit the panel and increase the cutting depth slightly to make a second pass *(right)*. Continue until the panel fits in the grooves.

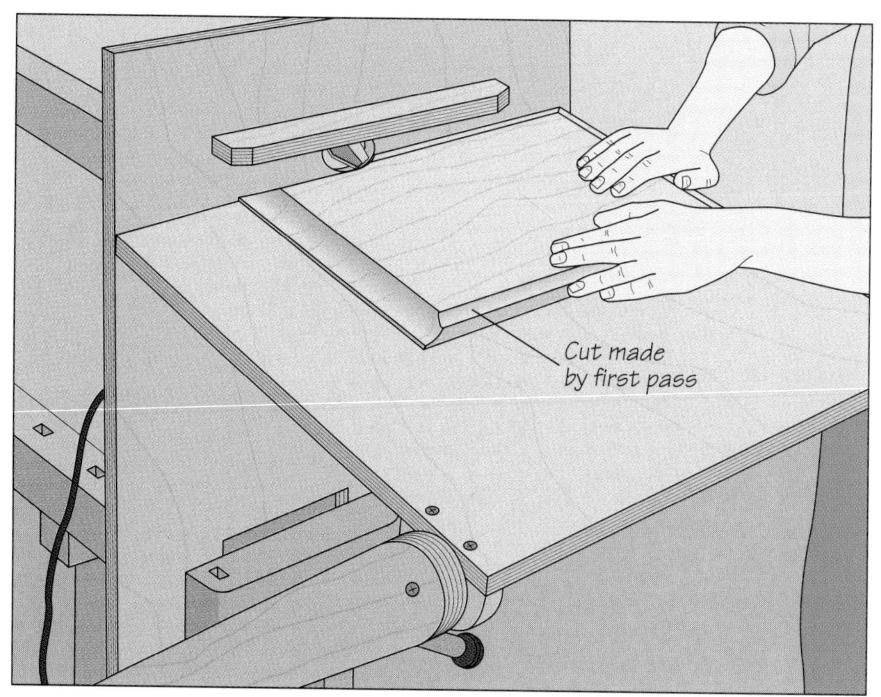

ADJUSTABLE ROUTING GUIDE

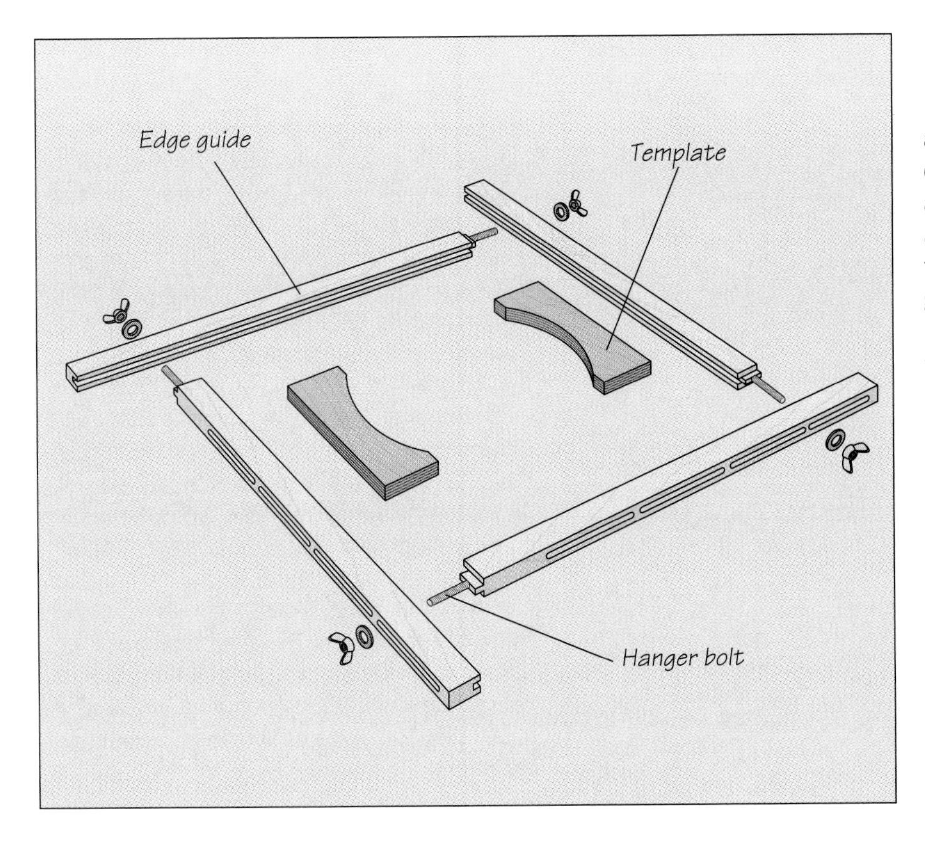

1 Building the jig
The jig shown at left is ideal for routing grooves in a rectangular pattern; it can also be fitted with templates for curved cuts. Saw the guides from 1-by-2 stock and rout a groove ⅜ inch deep and wide along the inside edge of each one. Cut a two-shouldered tenon at one end of each guide to fit in the grooves and bore a pilot hole into the middle of each tenon for a ⅜-inch-diameter hanger bolt. Screw the bolts in place, leaving enough thread protruding to feed it through the adjacent edge guide with a washer and wing nut. Finally, rout ⅜-inch-wide mortises through the guides; start about 3½ inches from the end with the tenon and make the mortises 4 inches long, separating them with about ½ inch of wood. Assemble the jig by slipping the tenons and bolts through the grooves and mortises of the adjacent guide and installing the washers and nuts. For curves, make templates like those shown in the illustration.

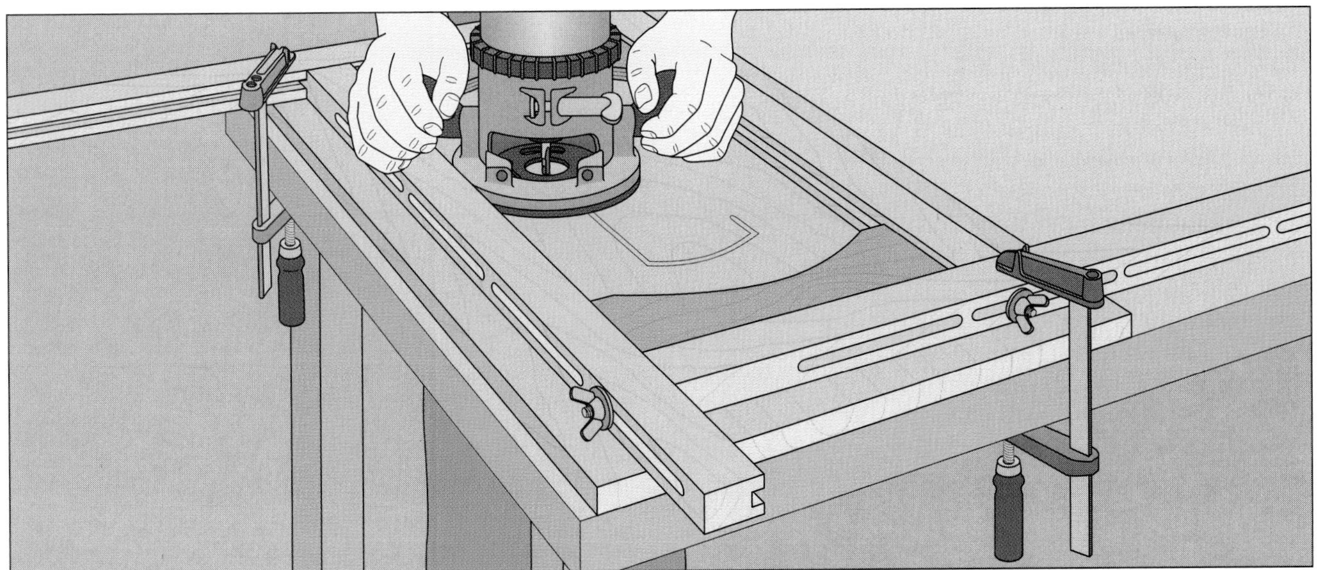

2 Routing the groove
Outline the pattern on your stock and lay it on a work surface. Loosen the wing nuts of the jig and position it on the stock so the edge guides frame the outline. Place the router flat on the workpiece and align the bit with one edge of the outline. Butt one of the edge guides flush against the router base plate. Repeat on the other edges until all guides and templates are in position. (Use double-sided tape to secure the templates to the workpiece.) Tighten the wing nuts and clamp the jig and workpiece to the table. After plunging the bit into the stock, make the cut in a clockwise direction, keeping the base plate flush against the edge guide or template at all times. For repeat cuts, simply clamp the jig to the new workpiece and rout the pattern *(above)*.

A SLIDING DOVETAIL JIG

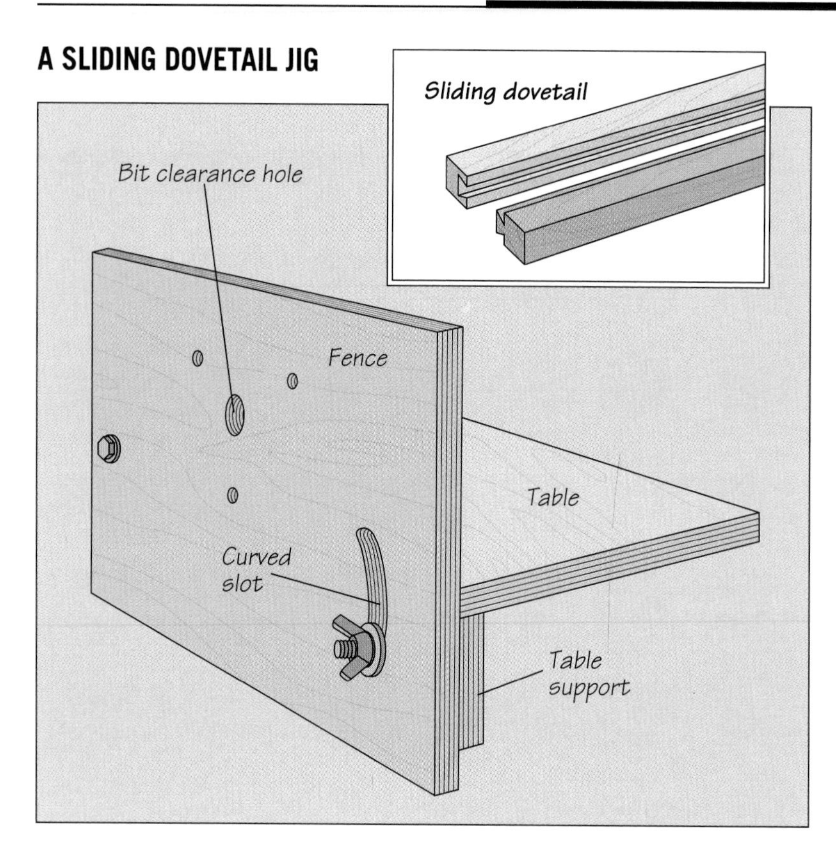

1 Building the jig
The jig shown at left allows you to rout sliding dovetails *(inset)* without a router table. Cut the fence, table, and support piece from ¾-inch plywood. Make all the boards 16 inches long; the fence and table should be about 10 inches wide and the support piece about 3 inches wide. Screw the table to the table support so they form an L. Position the table 4 inches from the bottom of the fence and bore two holes through opposite ends of the fence into the table support. Use a router with a straight bit to lengthen the hole on the outfeed side of the fence into a curved slot. Attach the table support to the fence with carriage bolts, washers, and wing nuts. Leave the bolt at the infeed end loose enough for the table to pivot when the slotted end is raised or lowered. Remove the sub-base from your router and use it as a template to mark the screw holes and bit clearance hole on the fence. The bottom edge of the clearance hole should line up with the top of the jig table when the table is level.

2 Routing the joint
Secure the fence in a vise and rout the dovetail groove first, then the matching slide. For the groove, start by installing a straight bit in the router, attaching the tool to the jig fence, and adjusting the cutting depth. Set the groove workpiece face-down on the table, butting its edge against the bit. Loosen the wing nut at the slotted end and adjust the table to center the bit on the edge of the stock, then tighten the nut. Secure the workpiece with three featherboards, clamping one to the table and the other two to the fence on both sides of the cutter. Make the straight cut, then complete the groove with a dovetail bit. For the slide, set your workpiece on the table and lower the table to produce a ⅛-inch-wide cut. Make a pass on both sides, finishing each cut with a push stick *(right)*. (In the illustration, the featherboard on the outfeed side of the fence has been removed for clarity.) Test-fit the joint. If necessary, raise the table slightly and make another pass on each side of the stock.

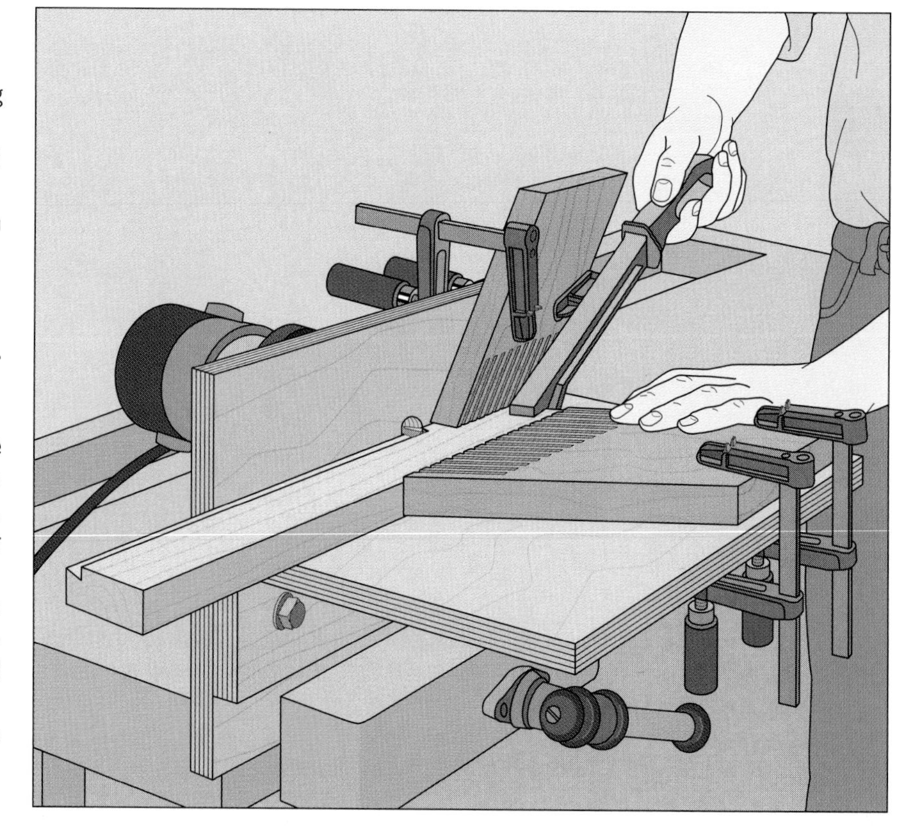

A CORNER HALF-LAP JIG

1 Building the jig
Cut the two base pieces and the stop block from plywood the same thickness as your stock. The base pieces should be wide enough to accommodate the edge and end guides and support your router's base plate. Use solid wood strips for the four guides. Next, mark the shoulder of the half-lap *(inset)* on one workpiece and butt the base pieces against its edges with the shoulder mark positioned near the middle of the boards. Install a straight bit in the router and align it with the shoulder mark, then mount an end guide across the base pieces and against the tool's base plate. Repeat the procedure to position a guide at the opposite end. Now align the bit with the edges of the workpiece and attach the edge guides, leaving a slight gap between the router base plate and each guide. (The first half-lap you make will rout reference notches in the base pieces.) Finally, install the stop block under one end guide, against the end of the workpiece. Countersink all fasteners.

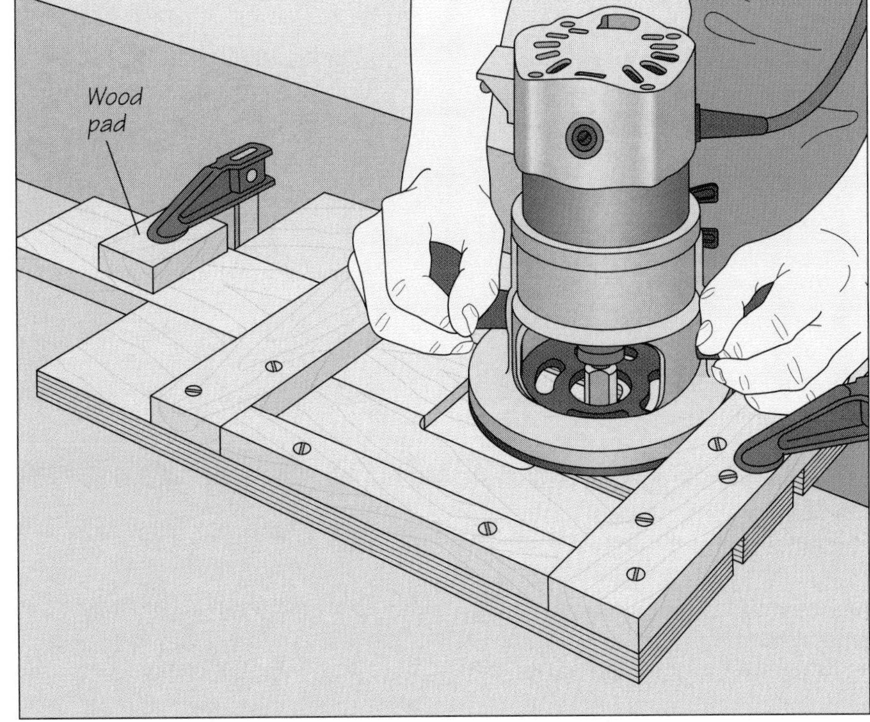

2 Routing the half-laps
Clamp the jig to a work surface and slide the workpiece between the base pieces until it butts against the stop block. Protecting the stock with a wood pad, clamp the workpiece in place. Adjust the router's cutting depth to one-half the stock thickness. Then, with the router positioned inside the guides, turn on the tool and lower the bit into the workpiece. Guide the router in a clockwise direction to cut the outside edges of the half-lap, keeping the base plate flush against a guide at all times. Then rout out the remaining waste, feeding the tool against the direction of bit rotation as much as possible *(right)*. To rout a T or cross half-lap, simply remove the stop block so the cut can be made at any point along the face of the stock and clamp the workpiece on both ends.

AN ADJUSTABLE MORTISING JIG

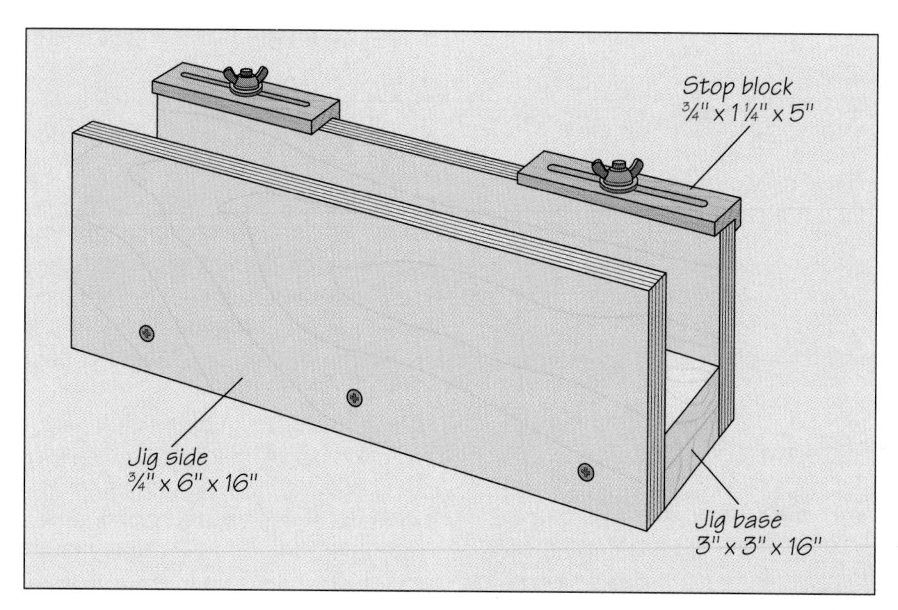

1 Making the jig
The jig shown at left will guide your router and secure the workpiece as you cut a mortise. The dimensions suggested in the illustration will suit most routers. Cut the jig sides from ¾-inch plywood. Make the base of laminated solid wood. Attach the sides to the base with countersunk screws, making sure the pieces are perfectly square to each other. Fashion each stop block from solid stock by cutting a rabbet ⅜ inch deep and 1 inch wide, then routing a 4-inch-long slot to accept a ¼-inch hanger bolt. Mount the bolts 3 inches from each end of one side, slip the stop blocks in place, and fix them with washers and wing nuts.

2 Routing a mortise
Set the workpiece on the jig's base with the mortise outline between the stop blocks and one surface flush against the side with the blocks. Place a shim under the stock so its top surface butts against the stop blocks, then clamp the workpiece to the jig and secure the jig in a workbench vise. Next, install a straight router bit the same diameter as the width of the mortise, set the depth of cut, and attach a commercial edge guide to the router base plate. Center the bit over the mortise outline and position the edge guide so it rests flush against the opposite side of the jig. Adjust each stop block by aligning the bit with the end of the mortise outline, butting the block against the router's base plate, and tightening the wing nut. Once the blocks are locked in position, turn on the tool with the bit clear of the workpiece. Gripping the router firmly, butt the edge guide against the jig, press the base plate against one stop block and plunge the bit into the workpiece. Hold the edge guide against the jig as you draw the router through the cut until it contacts the other stop block *(right)*. Cut a deep mortise in several passes, increasing the bit depth each time.

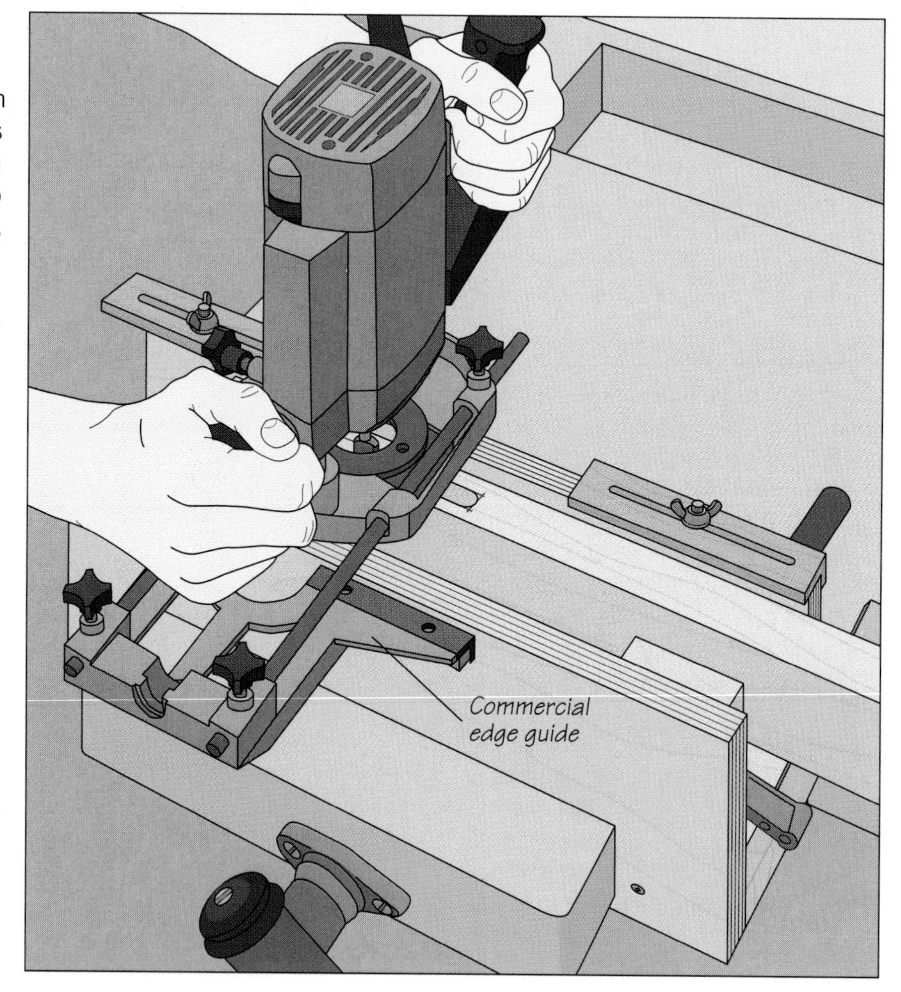

A MOVABLE-JAW MORTISING JIG

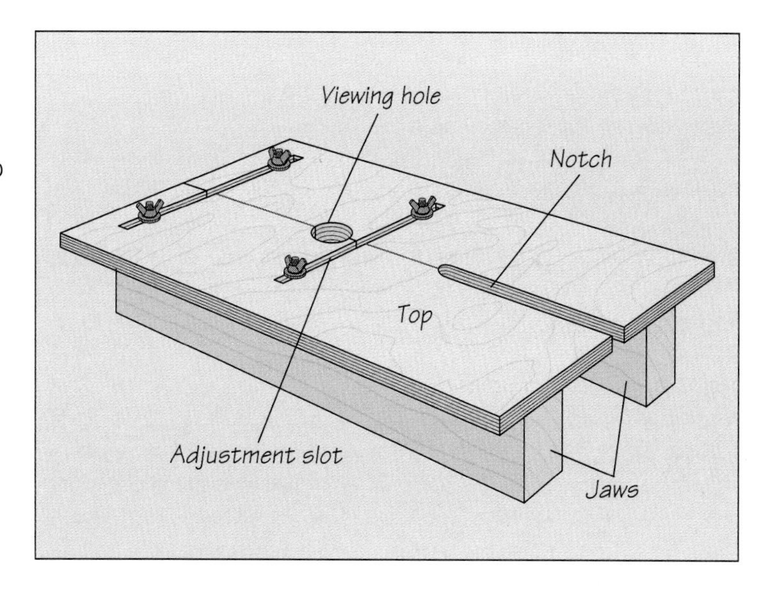

1 Making the jig
The jig at right allows you to rout perfectly centered mortises in stock of any thickness using a straight bit and a template guide. Cut the jig top from ¾-inch plywood; make the piece about 15 inches long and wide enough to accept the thickest board you expect to mortise. Cut the two jaws from 2-by-4-inch stock the same length as the top. To prepare the top, mark a line down its center and rout a notch centered over the line at one end. The notch should be the same width as the template guide you will use with your router bit, and long enough to accommodate the longest mortise you expect to cut. Next, rout two adjustment slots perpendicular to the centerline. Finally, bore a viewing hole between the two slots. To assemble the jig, screw hanger bolts into the jaws and fasten the top to the jaws with washers and wing nuts.

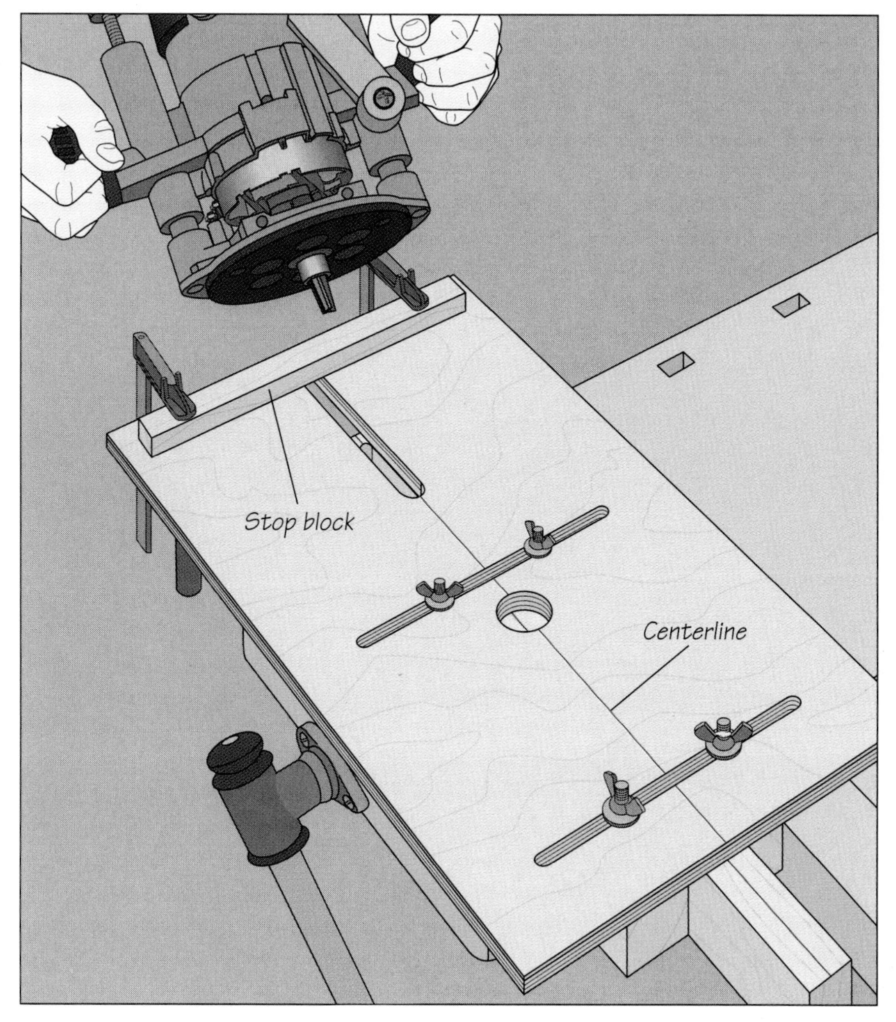

2 Cutting a mortise
Mark a line down the center of the mortise outline on the workpiece. Loosen the wing nuts and secure the stock between the jaws so the mortise centerline is aligned with that of the jig top; make sure the top edge of the workpiece is butted against the underside of the top. Also align one end of the mortise outline under the end of the notch—offset by the distance from the edge of the bit to the template guide's edge—then tighten the hanger bolts. Align the edge of the bit with the other end of the outline and clamp a stop block to the top flush against the router base plate. Rout the mortise *(left)*, starting the cut with the template guide butted against the end of the notch and stopping it when the base plate contacts the stop block at the other end.

A BOX JOINT JIG

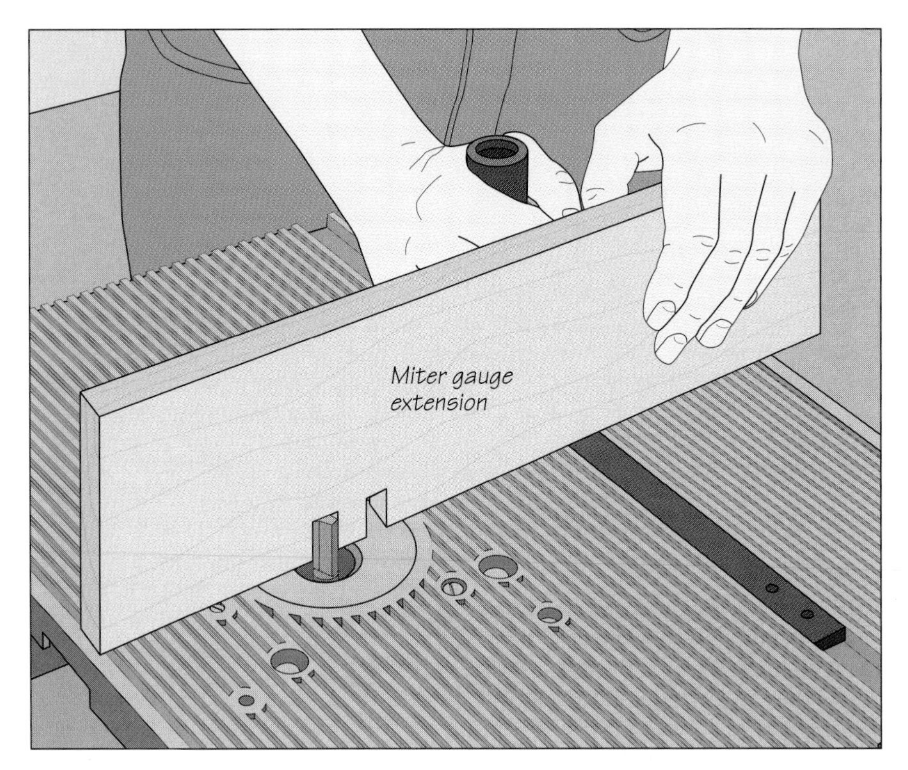

1 Setting up the jig
The jig shown on this page allows you to cut the notches for a box joint on a router table with little setup time. It consists simply of an extension board screwed to the miter gauge and fitted with a key that determines the spacing of the notches. Install a straight bit sized to the desired width of the notches and mount the router in a table. Set the depth of cut equal to the thickness of your stock and feed the extension into the bit to rout a notch through its bottom edge. Reposition the extension so that the gap between the notch and the bit equals the diameter of the bit, then screw it in place. Feed the extension into the bit again, cutting a second notch *(left).* Fashion a wood key to fit in the first notch and glue it in place so it projects about 1 inch from the extension board.

2 Cutting the box joint notches
Hold one edge of the workpiece against the key, butting its face against the miter gauge extension. Turn on the router and, hooking your thumbs around the gauge, slide the board into the bit, cutting the first notch *(right).* Fit the notch over the key and make a second cut. Continue cutting notches this way until you reach the opposite edge of the workpiece. To cut the notches in the mating end of the next board, fit the last notch of the first board over the key and butt one edge of the mating board against the first board. Move the entire assembly forward to cut the first notch in the mating board, holding both pieces flush against the miter gauge extension *(page 12).* Rout the remaining notches in the mating board the same way you made the cuts in the first board.

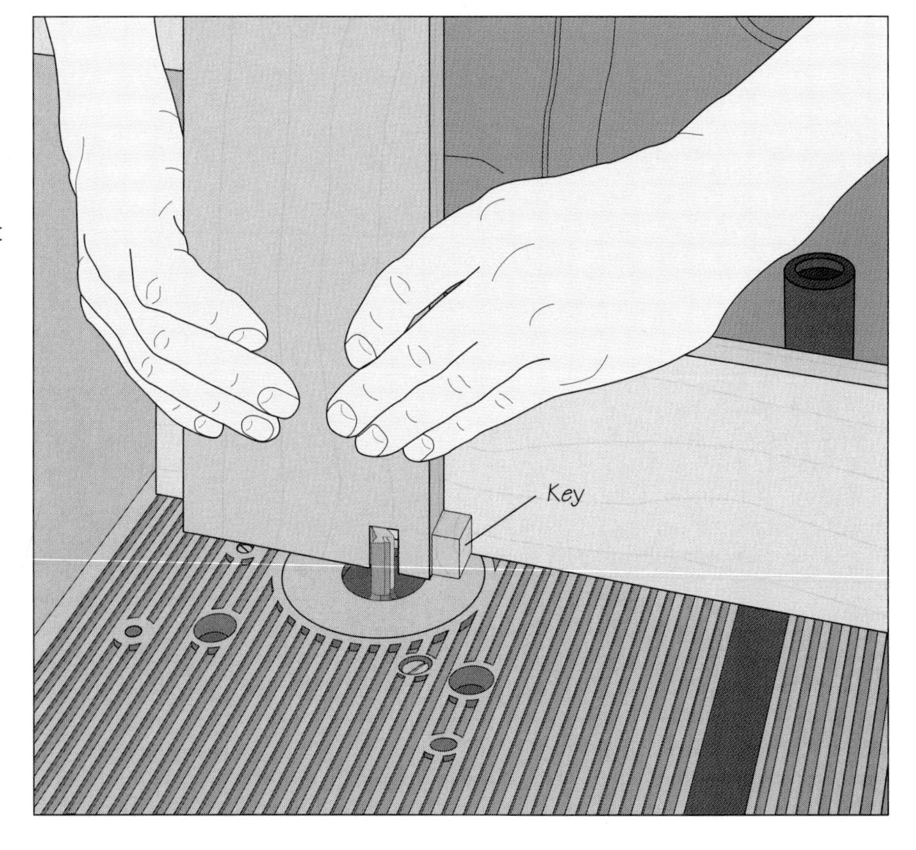

A CIRCLE-CUTTING JIG

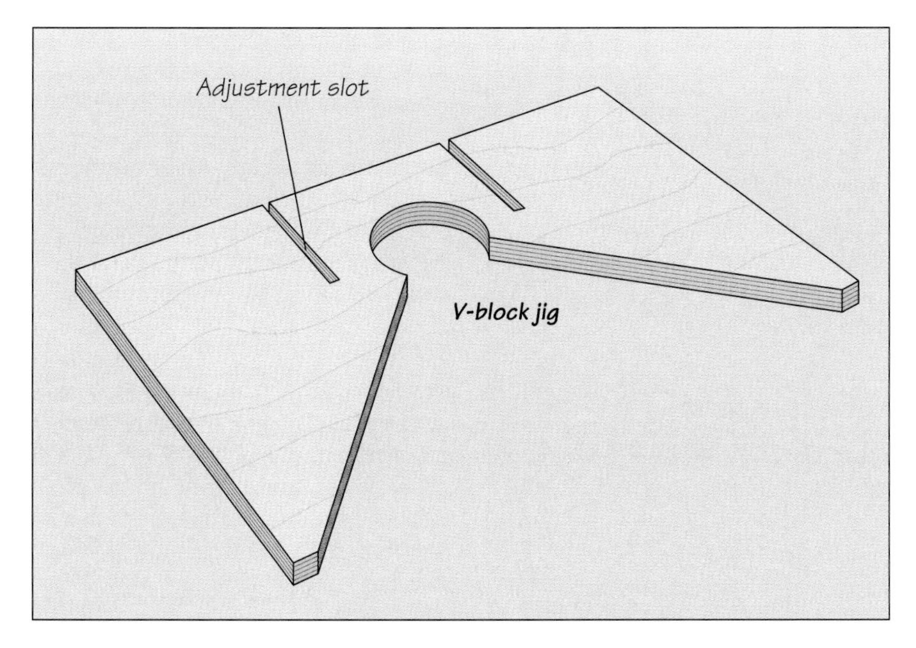

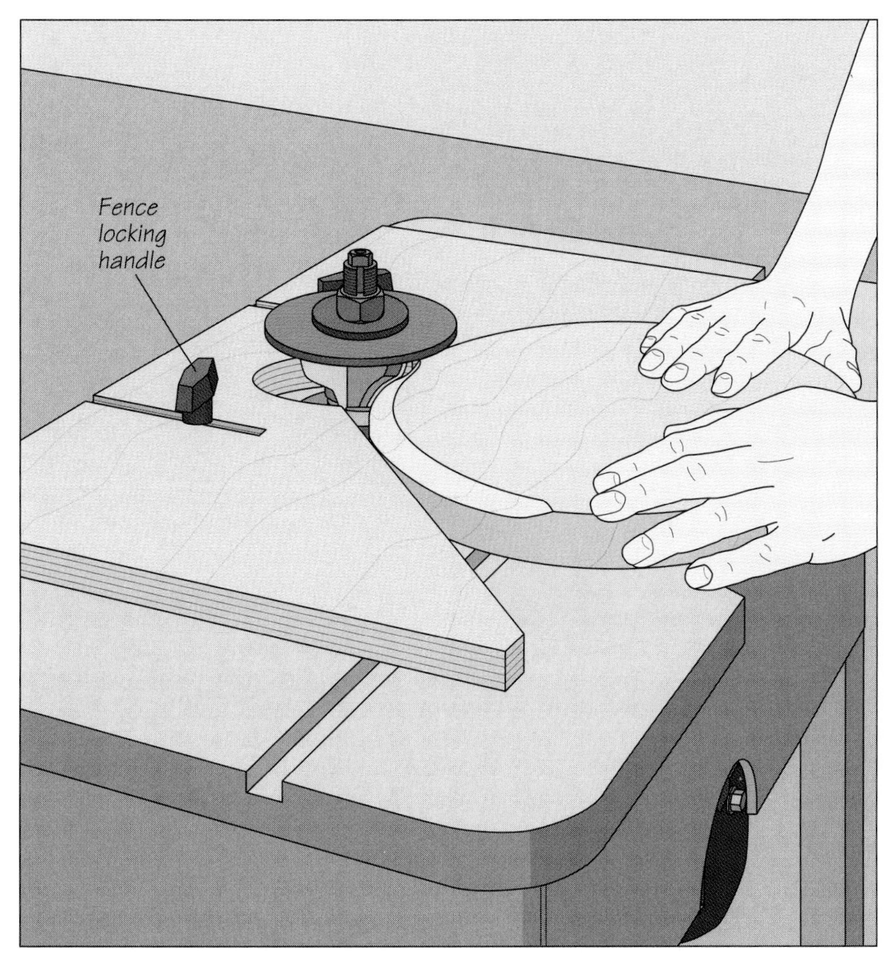

1 Making the jig
Shaping circular work freehand on the shaper is a risky job. One way to make the task safer and more precise is to use a V-block jig like the one shown at left. Build it from a piece of ¾-inch plywood about 14 inches wide and 24 inches long. To customize the jig for your shaper, hold it above the table flush with the back edge and mark the location of the spindle on the surface. Cut a right-angle wedge out of the jig, locating the apex of the angle at your marked point. Then cut a circle out of the jig centered on the apex; the hole should be large enough to accommodate the largest cutter you plan to use with the jig. Rout two adjustment slots into the back edge of the jig on either side of the hole—about ½ inch wide and 5 inches long. They must line up with the shaper's fence locking handles, as on the model shown. The jig can also be secured to the shaper by making it as long as the table and clamping it in place at either end.

2 Shaping circular work
Position the jig on the table, centering the bit in the hole. Seat the workpiece in the jig, butting it against both sides of the V, and adjust the jig and workpiece until the width of cut is set correctly. Secure the jig in place. You may want to make a test cut on a scrap piece the same thickness and diameter as your workpiece to be certain that the depth and width of cut are correct. Turn on the shaper and butt the workpiece against the outfeed side of the V. Slowly pivot the stock into the cutter until it rests firmly in the jig's V, moving it against the direction of cutter rotation to prevent kickback *(left)*. Continue rotating the workpiece until the entire circumference has been shaped, keeping the edge in contact with both sides of the jig throughout the cut.

AN EXTENDED SHAPER FEATHERBOARD

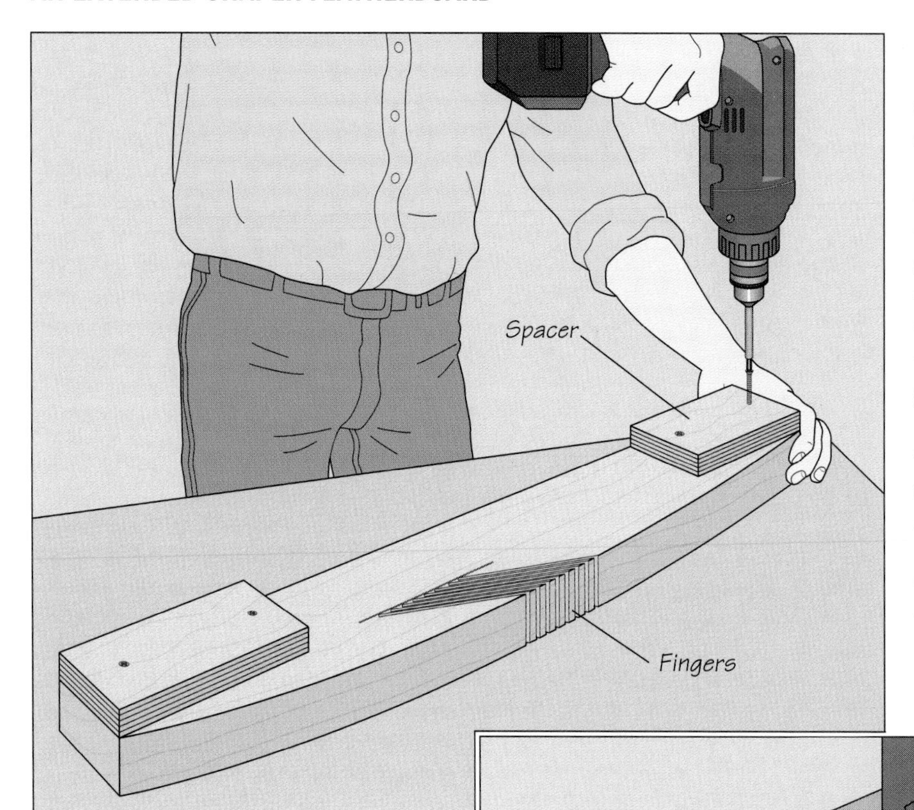

1 Making the featherboard
For wide cuts, such as shaping the edges of a panel, use an extra-wide featherboard like the one shown on this page. It will both press the panel against the table and shield your fingers from the cutter. Cut a 2-by-4 at least as long as your shaper's fence, set the board against the fence, and outline the location of the cutter on it. Curve the bottom edge of the featherboard slightly so that only the fingers will contact the panel during the shaping operation. Bandsaw a series of ¼-inch-wide slots at a shallow angle within the outline, creating a row of sturdy but pliable fingers. Screw two spacers to the back face of the featherboard so the jig will clear the cutter; countersink all fasteners *(left)*.

2 Raising the panel
Clamp the featherboard to the fence, centering the fingers over the bit, and turn on the shaper. For each pass, use your right hand to slowly feed the workpiece into the cutter; use your left hand to keep the panel against the fence *(right)*.

TWO SHAPER GUARDS

Building a fence-mounted guard

The shaper guard shown at right is ideal for fence-guided operations. Cut the pieces from ¾-inch plywood, making the guard in the shape of an arc large enough to extend from the fence and shield the cutter completely. The support board should be wide enough to be clamped to the fence when the guard is almost touching the spindle. Screw the guard flush with the bottom edge of the support board; countersink the fasteners. Next clamp the jig in position and mark a point on the guard above the cutter. Remove the jig and bore a 1¼-inch-diameter hole through the guard at the mark; the hole will allow you to view the cutter during shaping operations.

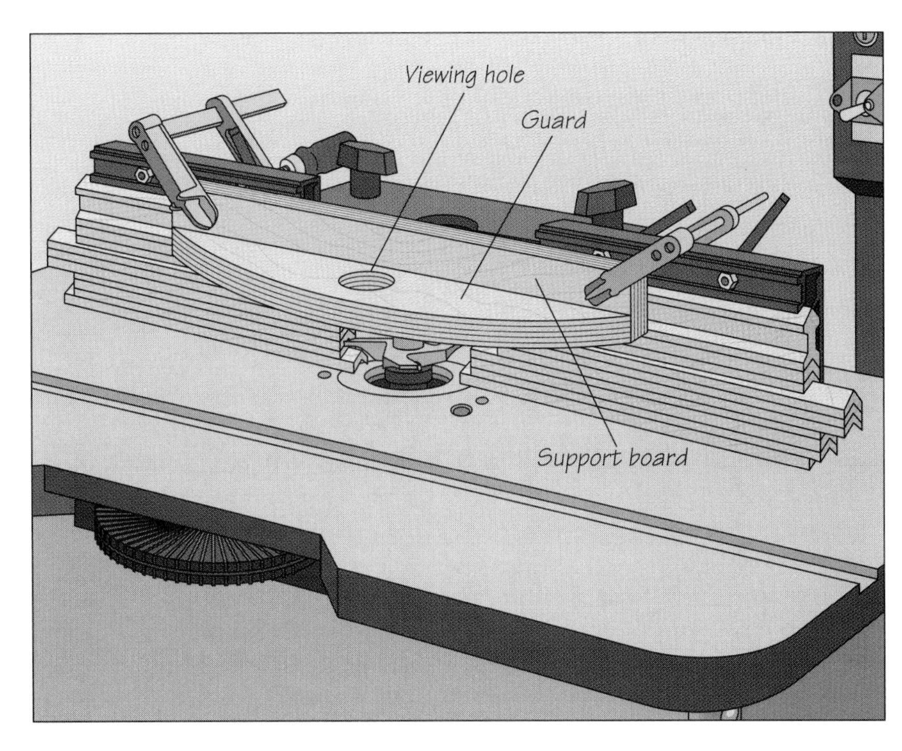

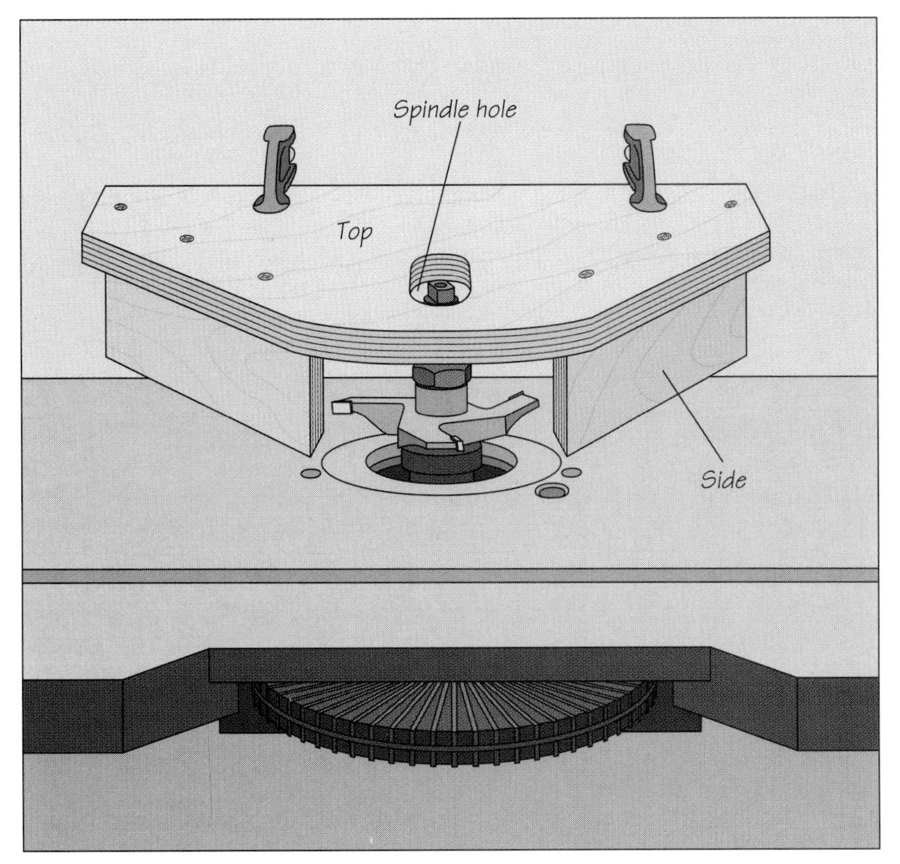

Making a freestanding guard

For freehand shaping, make a guard like the one shown at left. Sawn from ¾-inch plywood, it covers the cutter from the shaper's top, back, and sides. Cut the top about 16 inches long and wide enough to extend from the back of the table to about 1½ inches in front of the cutter. Bevel the front ends of the sides so they can be positioned as close as possible to the cutter. Rip the sides so the top will sit above the bit with just enough clearance for you to see the cutter. Hold the top on the table and mark a point on it directly over the spindle. Cut an oval-shaped hole through the top at the mark, large enough to clear the spindle and allow you to move the guard across the table slightly to accommodate different cutters. Fasten the top to the sides with countersunk screws. To use the guard, position it on the table with the spindle projecting through the top, and with the sides as close as possible to the cutting edges. Clamp the guard in place.

VACUUM JIGS

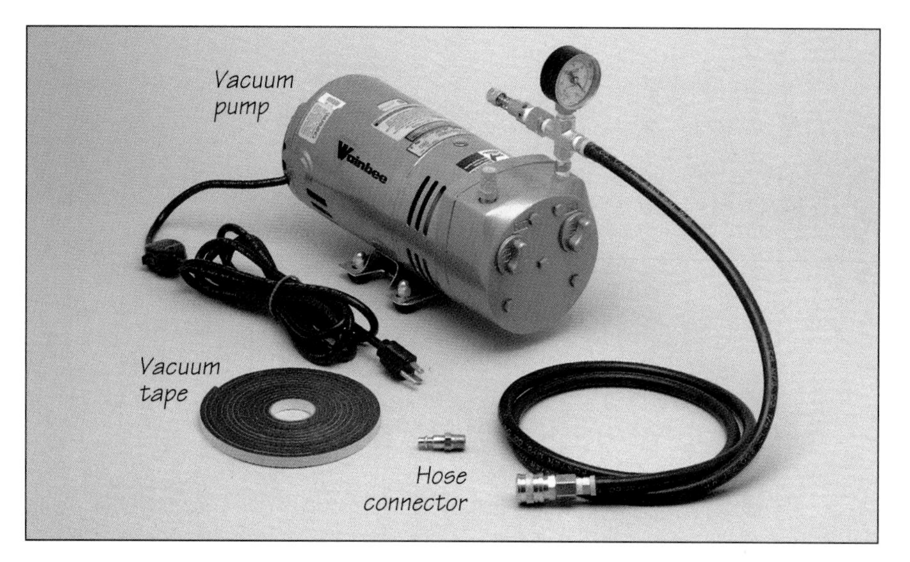

The heart of the vacuum system is the pump, here a 1/3 horsepower oil-less model, which draws air at a maximum of 4.5 cubic feet per minute. The hose features a quick coupler that attaches to a connector that is screwed into a hole through the template or featherboard. You will also need to use vacuum tape or closed-cell foam weatherstripping as a gasket to seal the cavity between template and workpiece or featherboard and work table.

The vacuum system shown here is an excellent way to anchor featherboards to work tables and fasten templates to workpieces. The system is more convenient than conventional clamping and offers as much holding power without risking damage to stock. The only limitation is that mating surfaces must be flat and smooth.

To set up a vacuum system, you need the parts shown in the photo at left. The tape is fastened to the underside of the featherboard or template, creating a cavity. The hose from the pump is inserted in a hole in the featherboard or template. When the jig is placed on the surface, the pump sucks the air from the cavity, producing a vacuum. Any pump rated at 3 cubic feet per minute or higher is adequate for the home workshop. If you own a compressor, you can convert it into a vacuum pump with a transducing pump.

A VACUUM FEATHERBOARD

Anchoring a featherboard to a saw table
Bore an outlet hole through the center of the featherboard. The hole's diameter should be slightly less than that of the threaded end of the hose connector you will use. Next, apply four strips of closed-cell vacuum tape to the underside of the featherboard, forming a quadrilateral with no gaps *(inset)*. Screw the hose connector into the outlet hole on the top face of the featherboard; use a wrench as shown opposite. To set up the vacuum jig, place the featherboard on the saw table—for the molding cut shown, it is positioned to press the workpiece against the fence. Make certain the tape strips are flat on the table. Snap the quick coupler at the end of the vacuum pump hose onto the hose connector and turn on the pump. Air pressure will anchor the featherboard to the table as you feed the workpiece through the cut *(right)*.

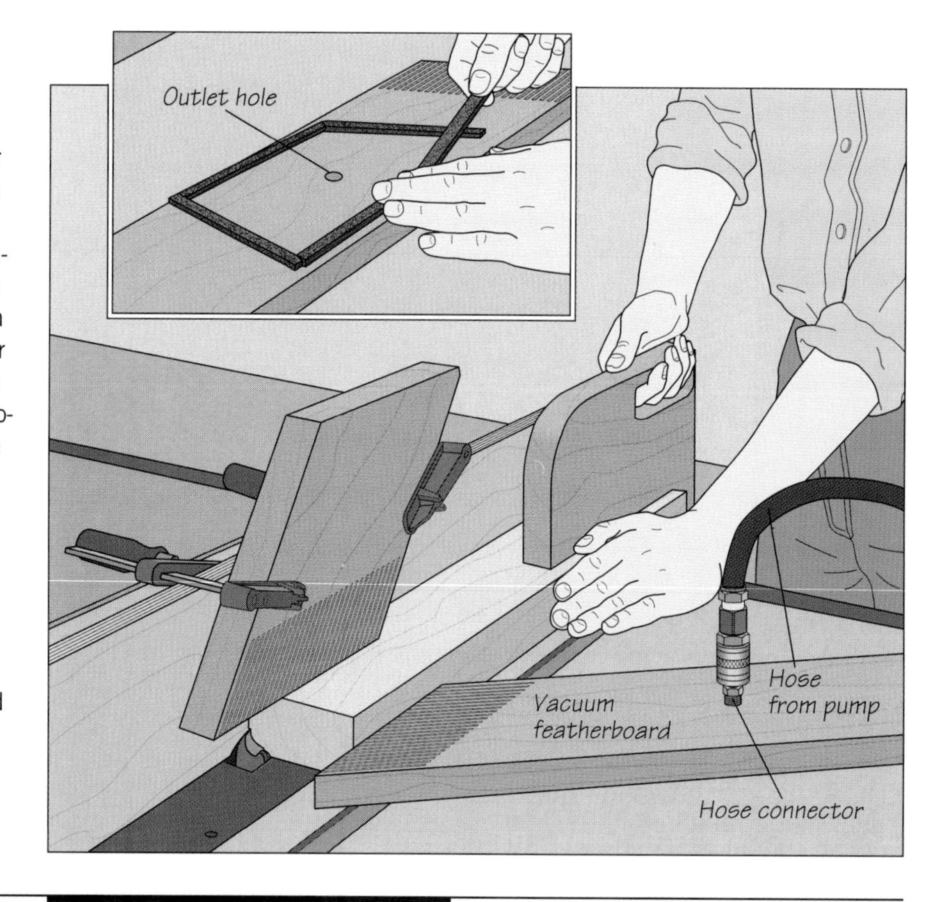

VACUUM TEMPLATE ROUTING

1 Installing the tape and connector

A vacuum provides an effective alternative to double-sided tape for fastening a plywood template atop a workpiece. Once your template is the proper size, trace its pattern on your stock and cut out most of the waste from your workpiece, leaving about ⅛ inch overhanging the template. Bore the outlet hole through the middle of the template and apply vacuum tape along the perimeter of its underside; make sure there are no gaps between adjacent pieces of tape. With thin stock, add two thin strips of tape on either side of the outlet hole to prevent the vacuum pressure from pulling the middle of the workpiece against the template *(inset)*. Attach the hose connector to the top of the template in the outlet hole *(right)*.

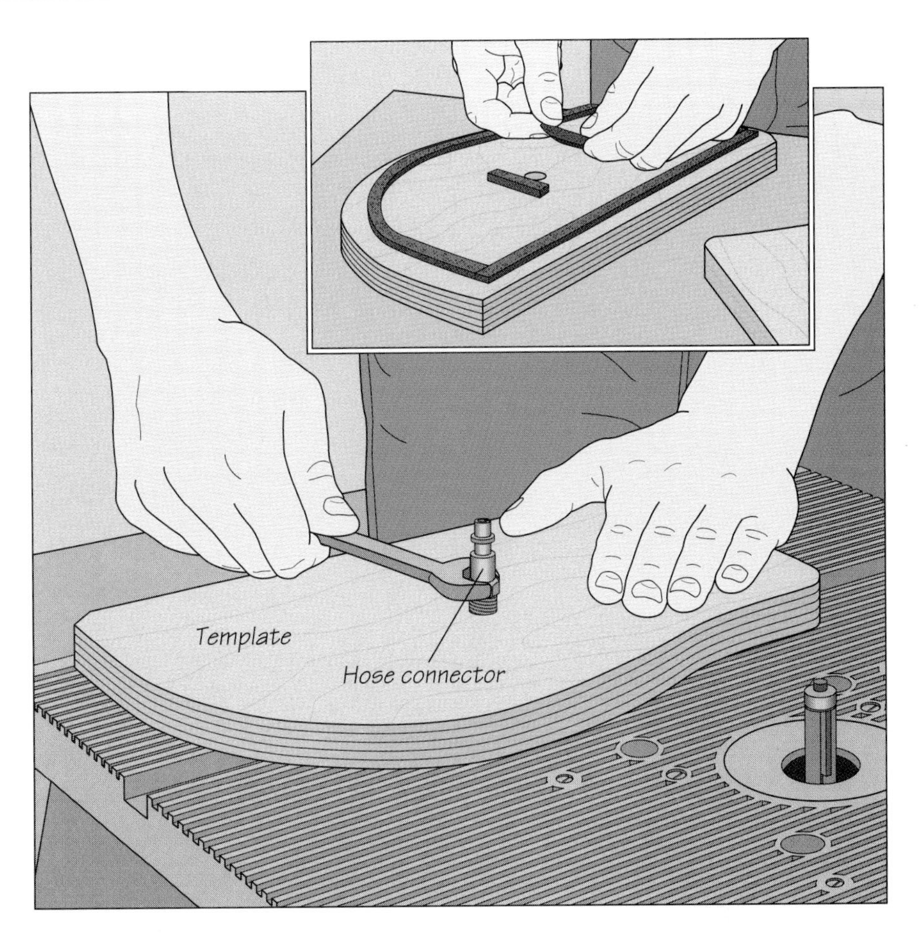

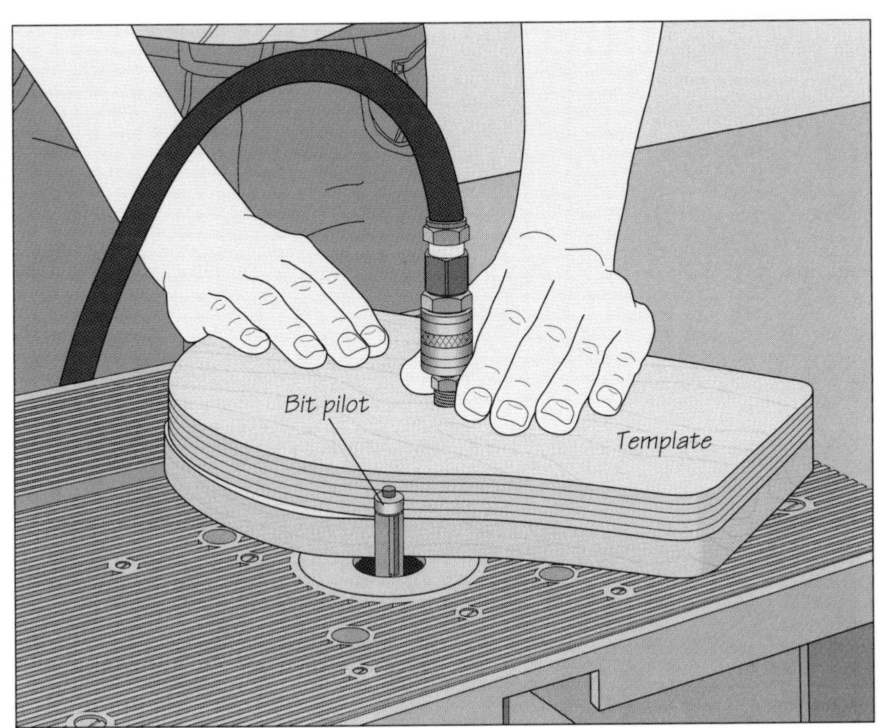

2 Routing the pattern

Install a piloted flush-trimming bit in a router, mount the tool in a table, and adjust the cutting height so the bit will shape the entire edge of the workpiece. Place the template tape-side-down centered on top of the workpiece. Attach the vacuum hose to the connector and switch on the pump to clamp the two boards together. Turn on the router and ease the stock into the bit until the template contacts the bit pilot *(left)*. Complete the cut, keeping the workpiece flat on the router table and the edge of the template pressed flush against the pilot; move against the direction of bit rotation.

CUTTING JIGS

From the time you cut rough lumber to length at the start of a project or miter trim to finish it, your power saws and handsaws are likely to be your most-used tools. Although many cutting tasks can be accomplished without them, the jigs shown in this chapter will make these operations easier—particularly when the same cut must be repeated on several workpieces.

With its intersecting arms, a table saw miter jig *(page 50)* guarantees miter joints that form perfect 90° angles. The tenoning jigs shown on pages 57 and 58 allow you to cut both parts of open mortise-and-tenon joints on the table saw.

Some jigs facilitate cutting tasks that are tough to perform freehand. The circle-cutting jigs for the saber saw *(page 43)* and band saw *(page 44)* help make quick work of circular tabletops. In tandem with your table saw, the raised panel jig *(page 52)* can produce beveled panels for frames.

These jigs will save you time in the shop. An added benefit is that most can be built from scrap wood, making them considerably less costly than store-bought counterparts.

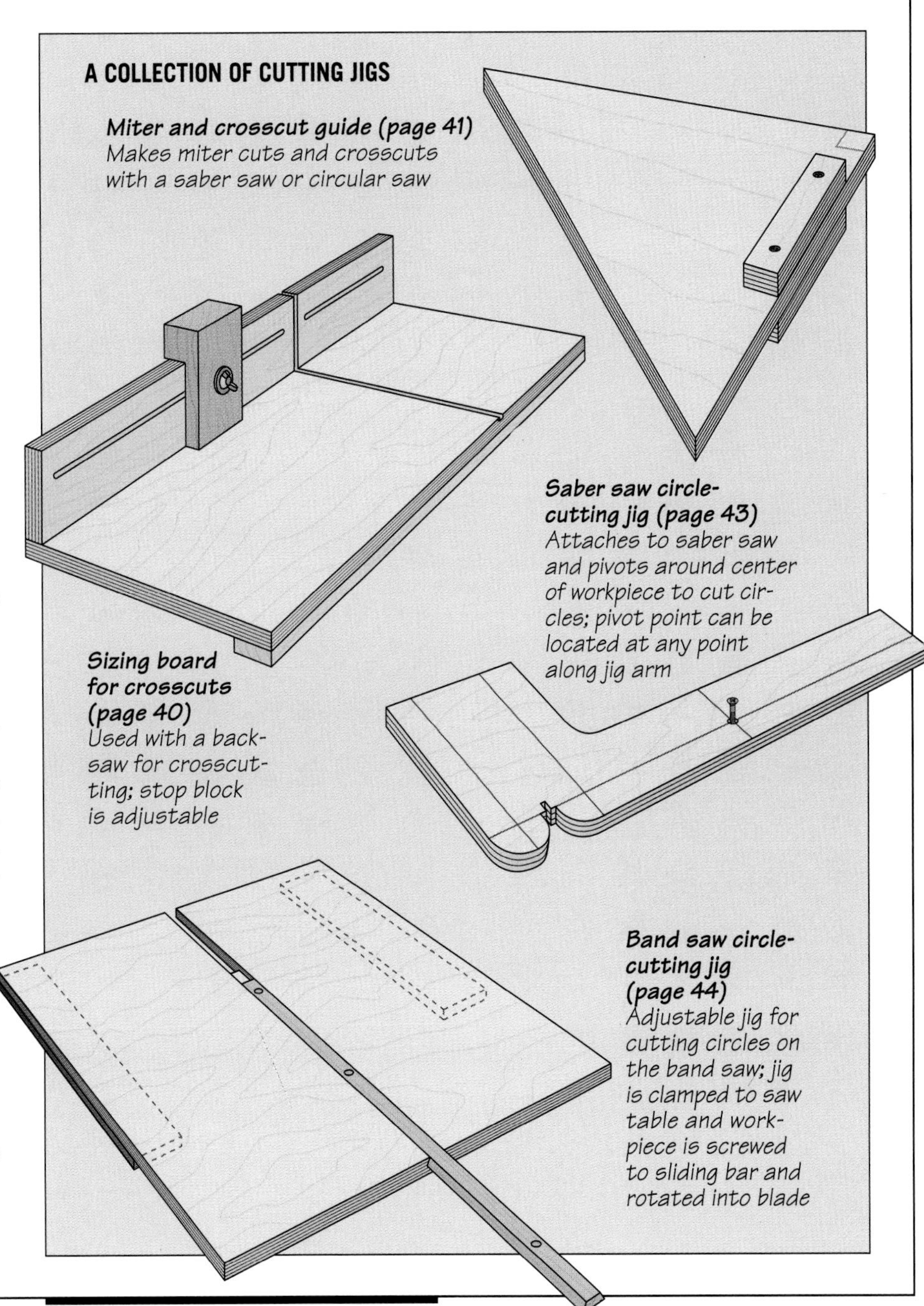

The crosscut jig shown at left provides a safe and accurate way to cut long, wide, or heavy stock on the table saw. Because it slides in the saw table's miter slots, this sturdy, adjustable jig makes it easy to hold workpieces square to the blade.

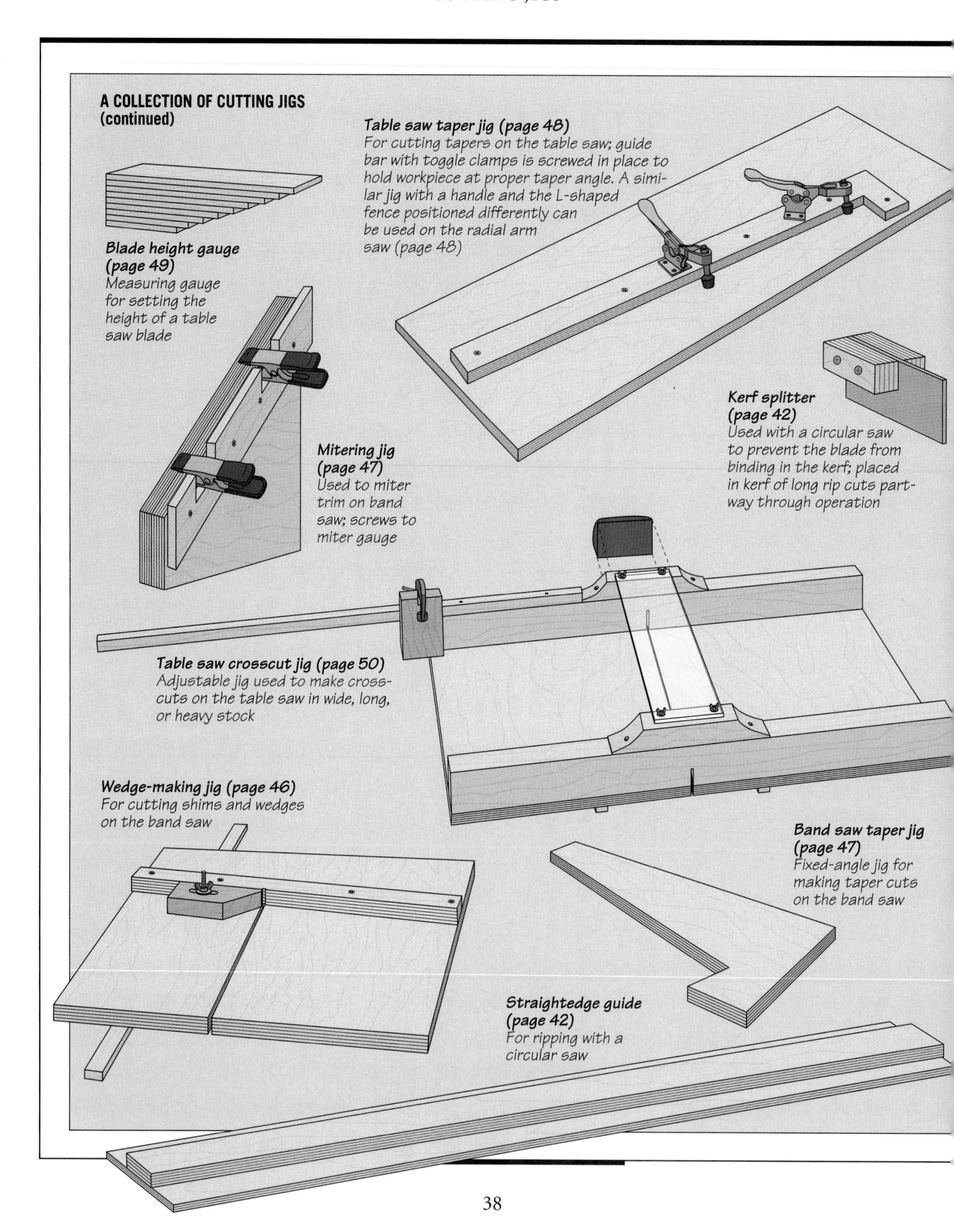
A COLLECTION OF CUTTING JIGS (continued)
Table saw taper jig (page 48)
For cutting tapers on the table saw; guide bar with toggle clamps is screwed in place to hold workpiece at proper taper angle. A similar jig with a handle and the L-shaped fence positioned differently can be used on the radial arm saw (page 48)
Blade height gauge (page 49)
Measuring gauge for setting the height of a table saw blade
Kerf splitter (page 42)
Used with a circular saw to prevent the blade from binding in the kerf; placed in kerf of long rip cuts partway through operation
Mitering jig (page 47)
Used to miter trim on band saw; screws to miter gauge
Table saw crosscut jig (page 50)
Adjustable jig used to make crosscuts on the table saw in wide, long, or heavy stock
Wedge-making jig (page 46)
For cutting shims and wedges on the band saw
Band saw taper jig (page 47)
Fixed-angle jig for making taper cuts on the band saw
Straightedge guide (page 42)
For ripping with a circular saw

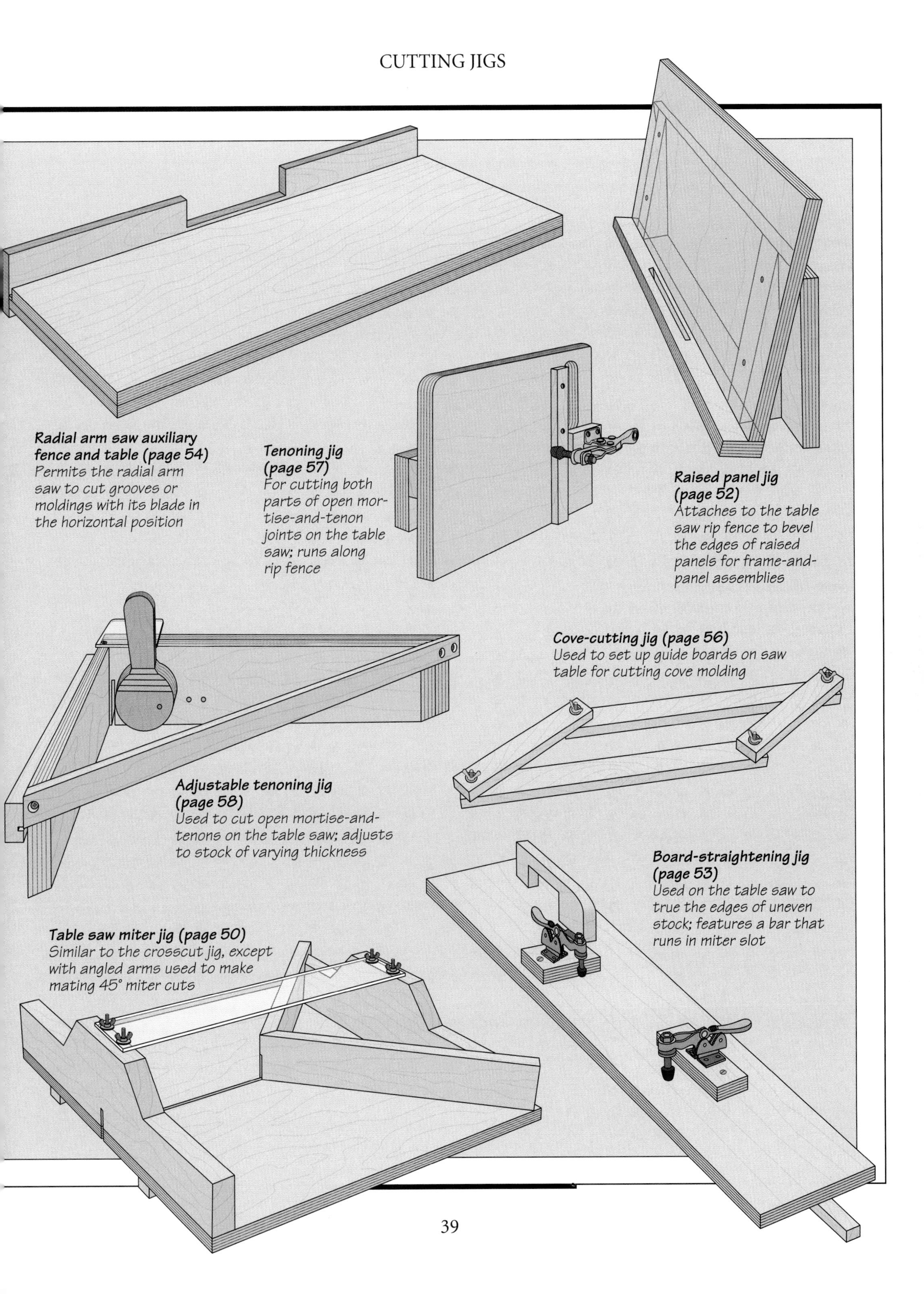
Radial arm saw auxiliary fence and table (page 54)
Permits the radial arm saw to cut grooves or moldings with its blade in the horizontal position
Tenoning jig (page 57)
For cutting both parts of open mortise-and-tenon joints on the table saw; runs along rip fence
Raised panel jig (page 52)
Attaches to the table saw rip fence to bevel the edges of raised panels for frame-and-panel assemblies
Cove-cutting jig (page 56)
Used to set up guide boards on saw table for cutting cove molding
Adjustable tenoning jig (page 58)
Used to cut open mortise-and-tenons on the table saw; adjusts to stock of varying thickness
Board-straightening jig (page 53)
Used on the table saw to true the edges of uneven stock; features a bar that runs in miter slot
Table saw miter jig (page 50)
Similar to the crosscut jig, except with angled arms used to make mating 45° miter cuts

SIZING BOARD FOR CROSSCUTS

1 Building the jig

The jig shown at right makes it easy to crosscut several workpieces to the same length by hand. Its adjustable stop block can be positioned at varying distances from the kerf in the fence. Cut the base and fence from ¾-inch plywood to the dimensions suggested in the illustration. Use solid wood for the stop block and lip. Screw the lip to the underside of the base, taking care to align the edges of the two pieces. Saw the fence into two segments about 7 inches from one end and use a router fitted with a ¼-inch bit to cut grooves through both pieces about 1 inch from their top edges; stop the grooves about 2 inches from the ends of each piece. Screw the two fence sections to the base, ensuring that the gap between the two pieces is wide enough to accommodate a saw blade. Saw a 90° kerf across the surface of the base in line with the kerf in the fence. To prepare the stop block, cut a 3-inch-long rabbet on one face and bore a clearance hole through its center for a 1½-inch-long, ¼-inch-diameter carriage bolt. Fasten the block to the fence with the bolt, washer, and wing nut.

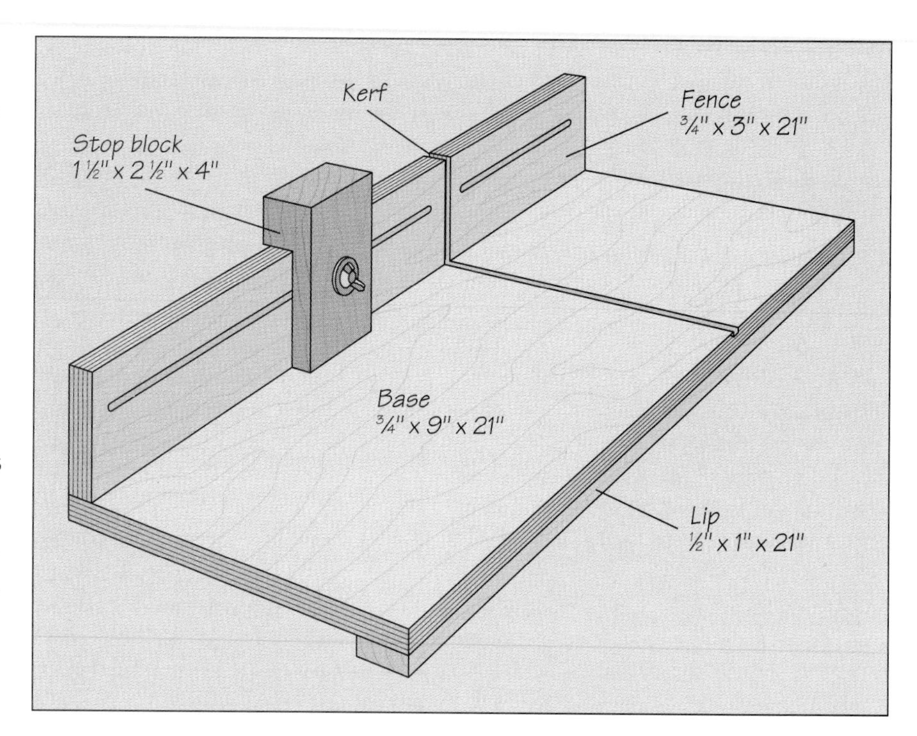

2 Making a crosscut

Butt the lip against the edge of your workbench, loosen the wing nut, and slide the stop block along the fence to the proper distance from the kerf between the two fence sections. Tighten the wing nut and butt the end of the workpiece against the stop block. Hold the stock firmly against the fence as you saw *(right)*.

MITER AND CROSSCUT GUIDE

1 Assembling the jig
The multipurpose edge guide shown at right will allow you to cut either 45° miter cuts or 90° crosscuts with a saber saw or a circular saw. Make the jig from a piece of ¾-inch plywood, referring to the illustration for suggested dimensions. Cut the base in the shape of a triangle with one 90° angle and two 45° angles. (To make a jig for 30° or 60° miter cuts, the sides should be 12, 16, and 20 inches or a variation of the 3-4-5 ratio.) Screw the fences to the base—one on each side—opposite one of the 45° angles. The fences must be flush with the edge of the jig base.

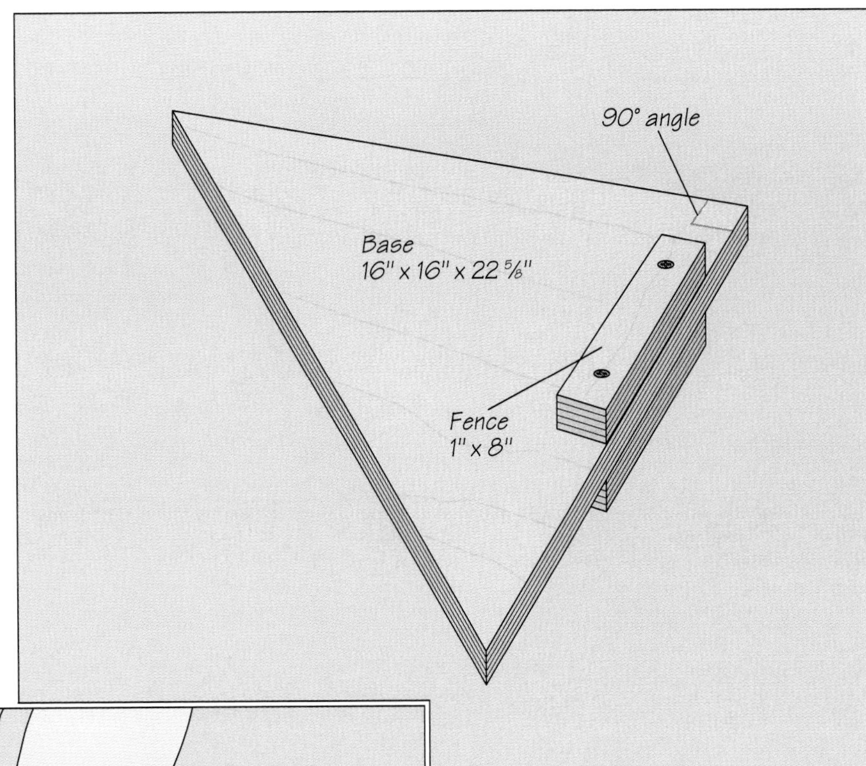

2 Making a miter cut
To cut a miter using the jig, set the stock on a work surface with the cutting line on the board extending off the table. Align the cutting edge with the line and butt the angled side of the jig against the saw's base plate, with the fence on the bottom of the guide flush against the edge of the workpiece. Clamp the jig in place and make the cut, keeping the saw flush against the jig throughout the operation *(left)*. To make a 90° crosscut, use the square side of the jig as your guide.

TWO CIRCULAR SAW JIGS

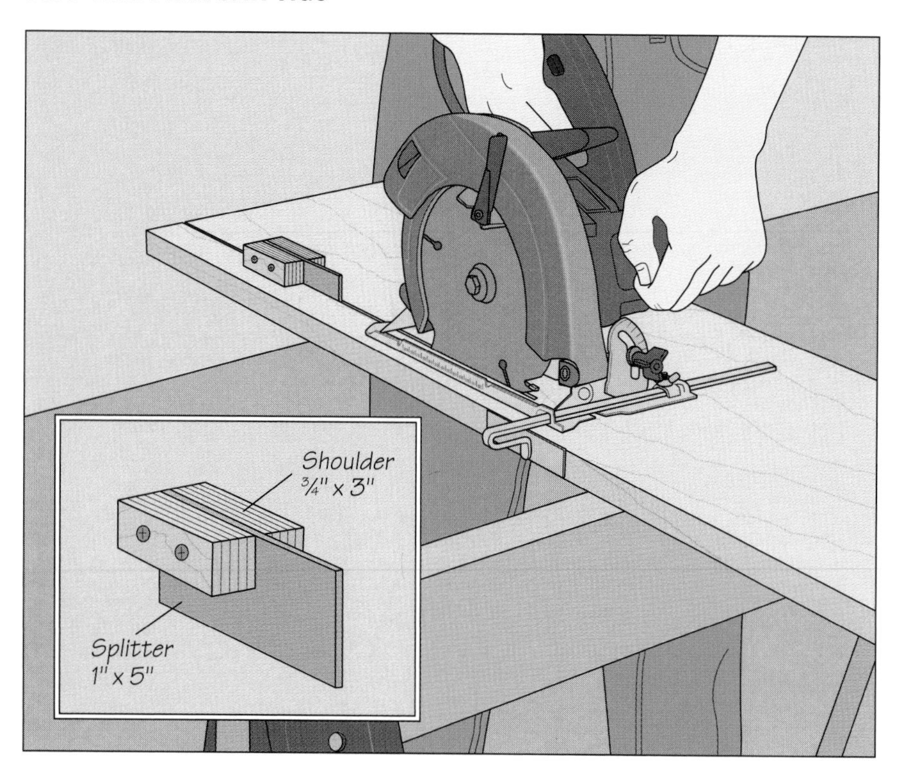

Using a kerf splitter
A kerf splitter like the one shown at left will help prevent a circular saw blade from binding in its kerf and kicking back. Choose ⅛-inch hardboard for the splitter piece and ¾-inch plywood for the shoulders; refer to the illustration for suggested dimensions. Fasten the three pieces together with screws. To use the jig, start the cut, turn off the saw, and insert the splitter in the kerf a few inches behind the saw. Back up the saw slightly, then continue the operation *(left)*. For particularly long cuts, advance the kerf splitter periodically to keep it near the saw.

Ripping with a straightedge guide
The jig shown above enables you to make accurate rip cuts in long manufactured panels like plywood. Make the base from ¼-inch plywood; use ¾-inch plywood for the edge strip. Glue the strip parallel to the base, offsetting its edge about 4 inches from one edge of the base. Trim the base to its proper width for your saw by butting the tool's base plate against the jig's edge strip and cutting along the length of the base. To use the jig, mark a cutting line on the panel and clamp the stock to a platform of 2-by-4s resting atop sawhorses. Clamp the guide to the panel, aligning the trimmed edge of the base with the cutting line on the workpiece. Make the cut *(above)*, keeping the saw's base plate flush against the edge strip throughout the operation.

TWO CIRCLE-CUTTING JIGS

CUTTING CIRCLES WITH THE SABER SAW

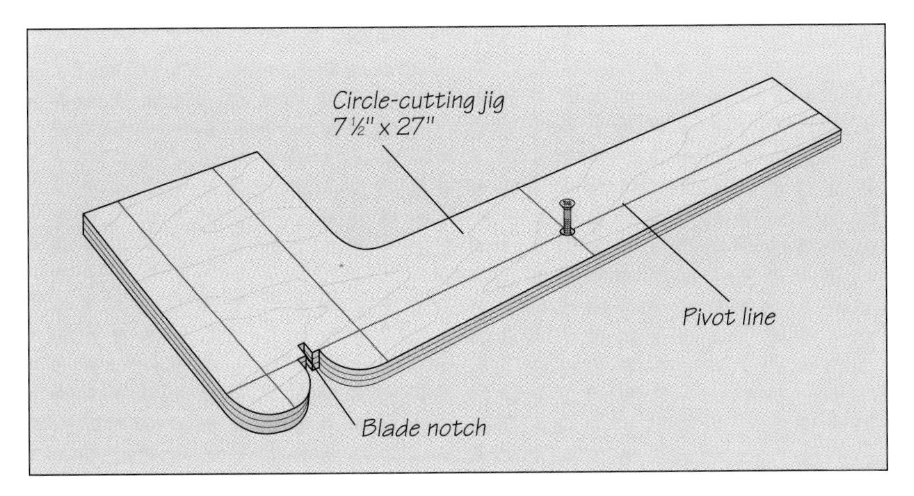

1 Building the jig
To cut circles bigger than the capacity of commercial saber saw jigs, use a shop-made guide customized for your saw. The exact size of the jig can vary, but the dimensions suggested in the illustration at left will yield a jig large enough to cut a circle to the edges of a 4-by-8 panel. Begin by removing the blade from your saw and outlining its base plate on a piece of ½-inch plywood. Reinstall the blade and cut along the marks, making the section that will be beneath the base plate slightly larger than the plate. Lighten the jig by trimming it to the shape of an L, then cut out the notch for the blade. Screw the jig to the base plate, ensuring that the back of the blade is flush against the bottom of the notch. Use a pencil to mark a pivot line on the jig that is aligned with the teeth of the blade.

2 Using the jig
Clamp down the stock with as much of the workpiece as possible extending off the table, using wood pads to protect the stock. Cut into the stock to bring the blade up to the outline of the circle you will be cutting. Then drive a screw into the jig on the pivot line at the center of the circle. Holding the saw and the stock firmly, cut out the circle *(below)*, shifting the clamps and workpiece as necessary.

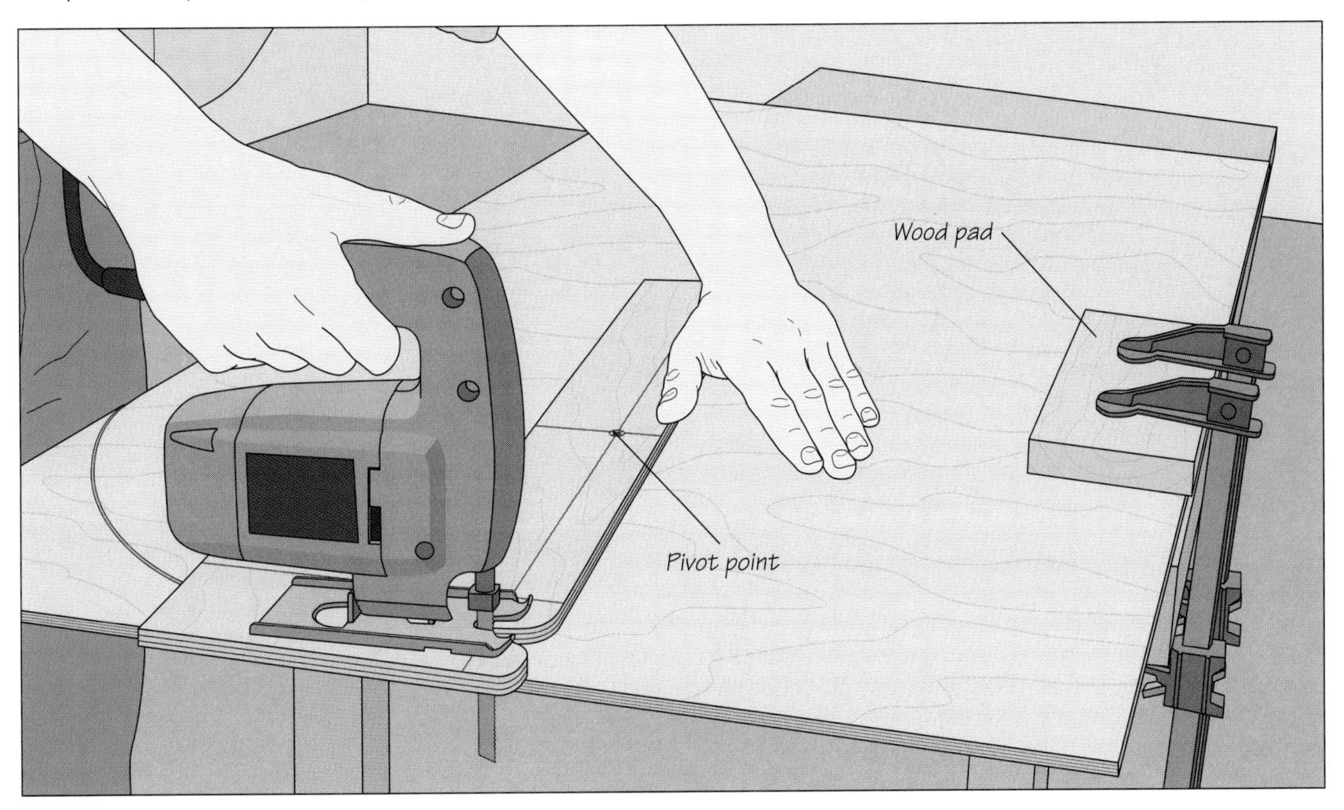

CUTTING CIRCLES ON THE BAND SAW

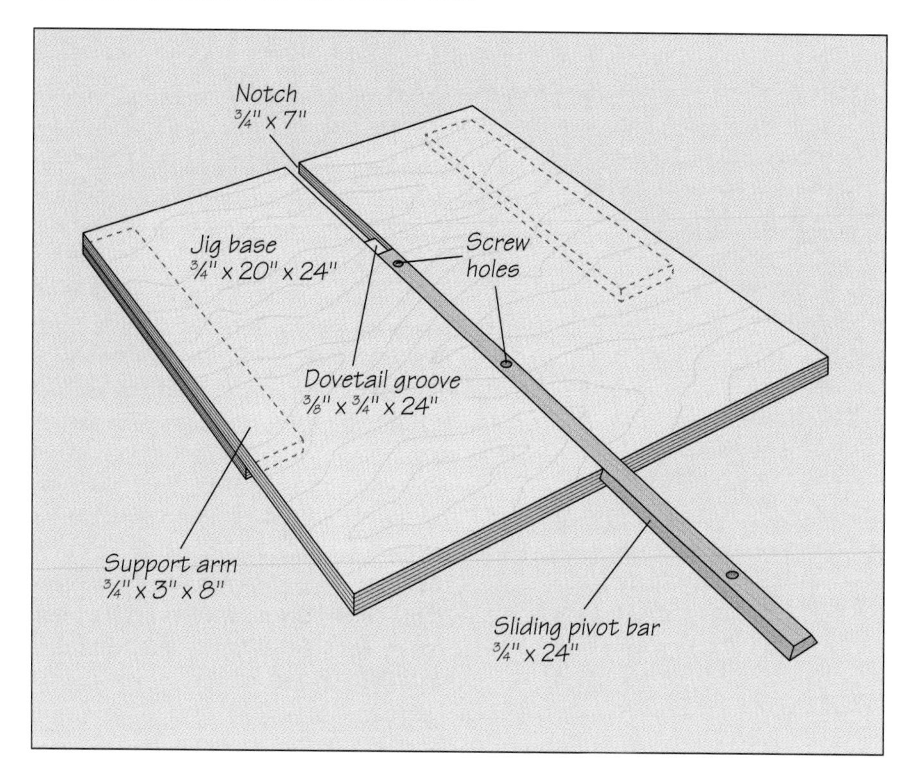

1 Building the jig
For cutting perfect circles on the band saw, use a circle-cutting jig custom-built for your tool like the one shown at left. Refer to the illustration for suggested dimensions. Use a router fitted with a dovetail bit to cut a ⅜-inch-deep groove in the middle of the jig base. Then use a table saw to rip a thin, beveled board that will slide smoothly in the channel. (Set the saw blade bevel angle by measuring the angle of the channel edges.) Cut out the notch on the band saw. Then position the jig base on the saw table so that the blade lies in the notch and the dovetail groove is perpendicular to the direction of cut. Now screw the support arms to the underside of the jig base; the arms should hug the sides of the band saw table. Bore two screw holes through the bottom of the dovetail channel in the jig base roughly 1 inch and 3 inches from the unnotched end; also bore three holes through the bar.

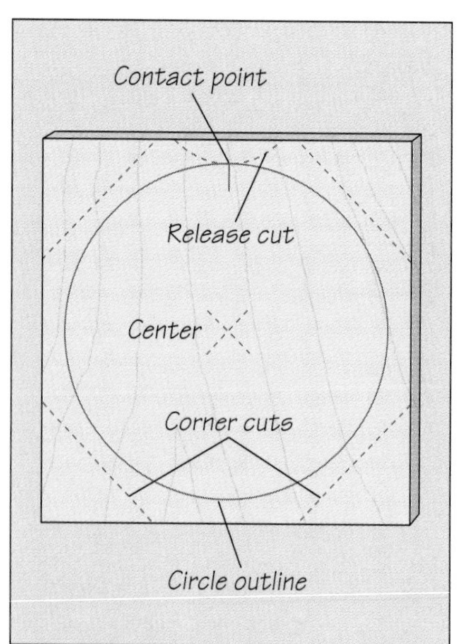

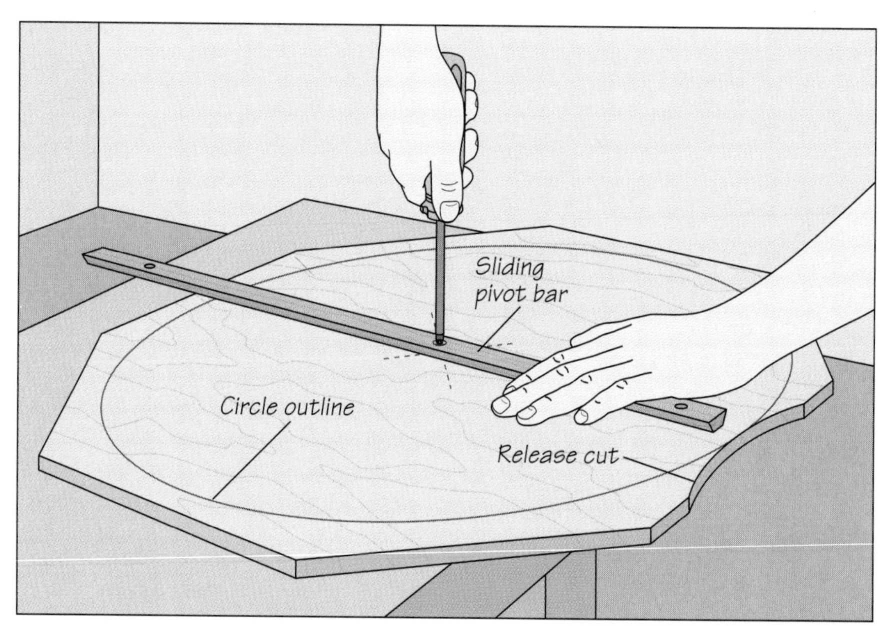

2 Preparing the workpiece
Mark the circumference and center of the circle you plan to cut on its underside. Then, use the band saw to cut off the four corners of the workpiece to keep it from hitting the clamps that will secure the jig to the table as the workpiece turns. Make a release cut from the edge of the workpiece to the marked circumference and veer off to the edge *(above, left)*. Screw the pivot bar to the center of the workpiece through one of the bar's holes *(above, right)*, leaving the screw loose enough to pivot the workpiece. Turn the workpiece over and mark the contact point where the blade touched the circumference during the release cut.

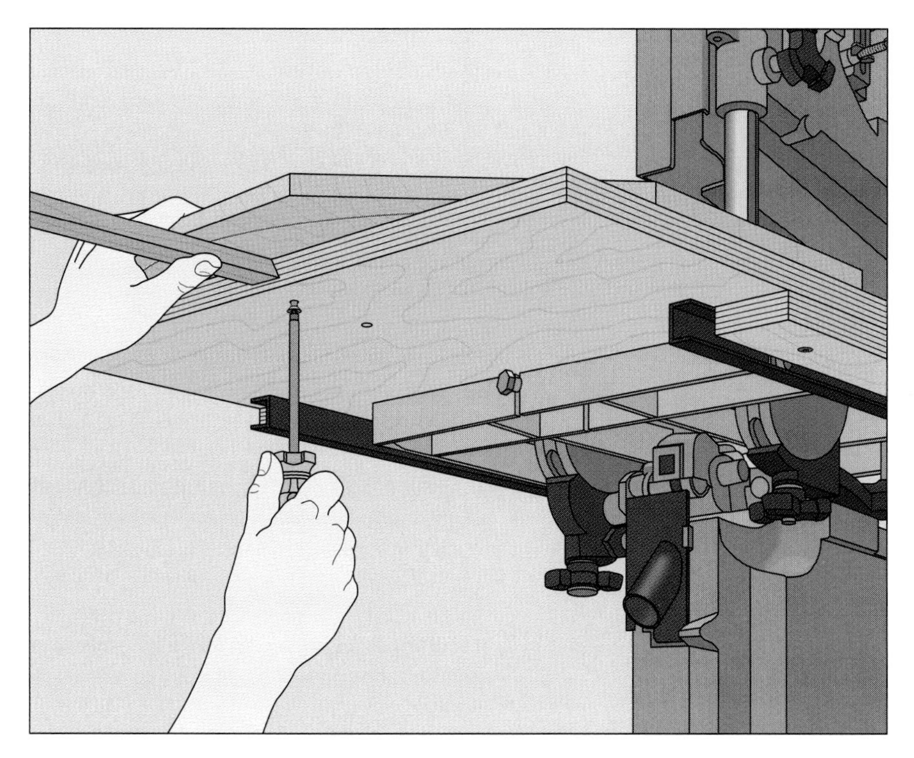

3 Securing the workpiece to the jig
Clamp the jig base to the band saw table, making sure the support arms are butted against the table's edges. Slide the pivot bar into the channel in the base and pivot the workpiece until the marked contact point touches the blade. Screw through one of the holes in the jig base to lock the pivot bar in place *(left)*.

4 Completing the circle
Turn on the saw and pivot the workpiece into the blade in a clockwise direction *(below)*, feeding the piece with your right hand until the cut is completed.

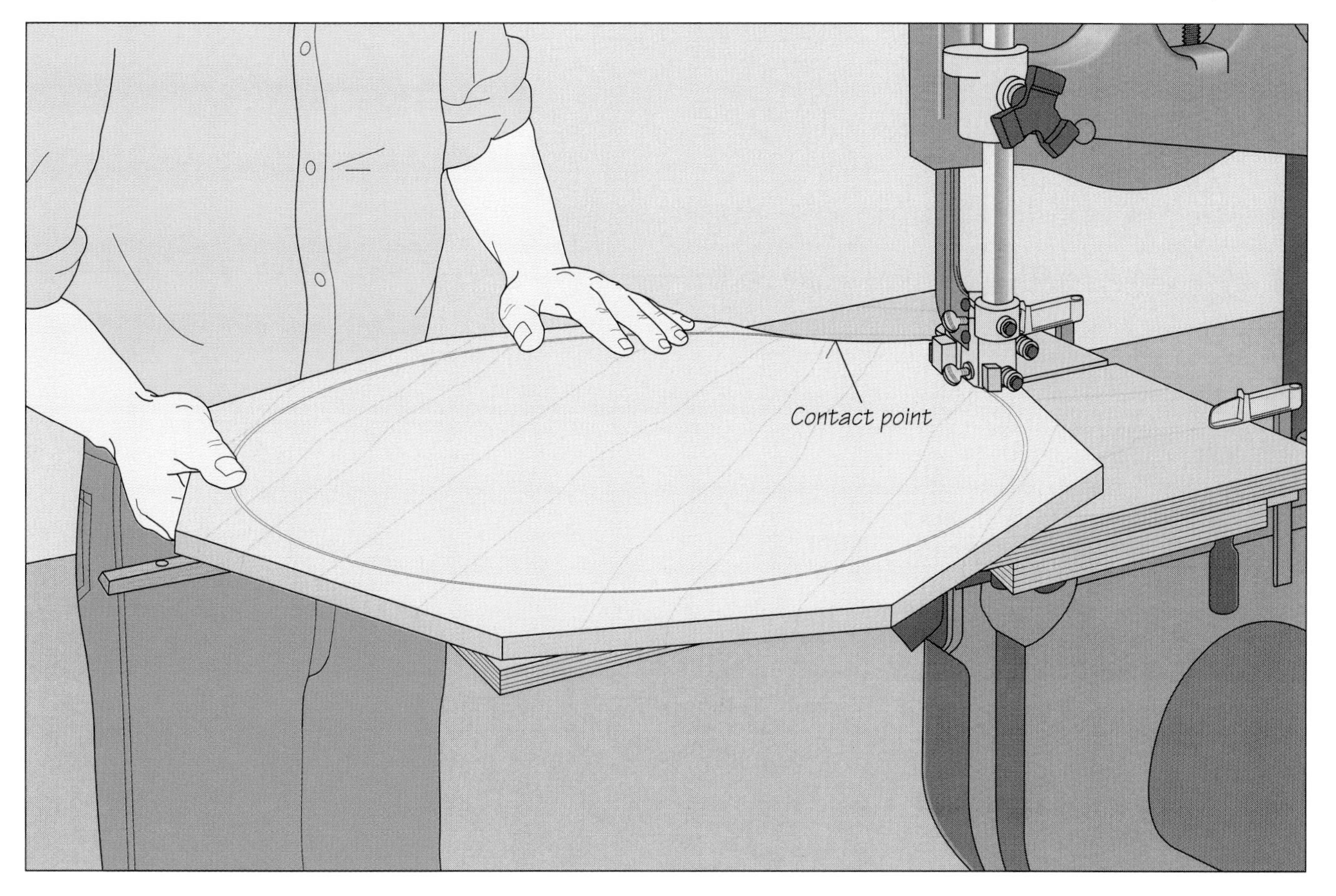

WEDGE-MAKING JIG

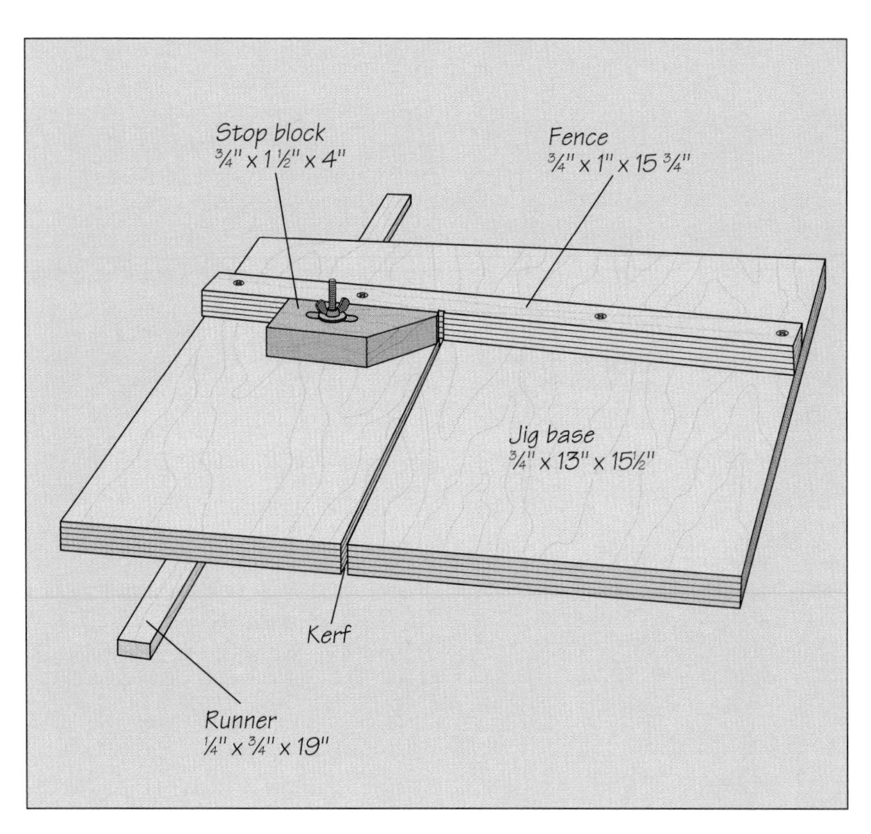

1 Building the jig
Small wedges are used for wedged tenons, or to shim cabinets on uneven floors. The jig shown at left allows you to make them quickly on the band saw. (You can also use the same device on a table saw.) Refer to the illustration for suggested dimensions, making sure the hardwood runner fits snugly in the saw table miter slot. Screw the runner to the underside of the base so that the runner extends beyond the tabletop and the base sits squarely on the table when the runner is in the slot; countersink the fasteners. Next, screw the fence to the top of the base; angle the fence at about 4° to the front and back edges of the base. Set the jig on the table with the runner in the slot, turn on the saw, and cut through the base until the blade contacts the fence. Turn off the saw, remove the jig, and cut a slot through the stop block for a machine bolt. Attach the block to the base, adding a washer and wing nut. The block should be flush against the fence with the tip of its angled end aligned with the kerf.

2 Cutting wedges
For your wedge stock, cut a strip of cross-grain wood from the end of a board; make it as wide as the desired length of the wedges. Position the jig on the saw table. Holding your stock with its edge flush against the fence and one end butted against the stop block, feed the jig across the table. Make sure your hands are clear of the blade as you cut each wedge *(right)*. To create 4° angle wedges, square the end of your stock on the table saw before each cut. If you simply flip the workpiece between cuts on the band saw, all the wedges after the first will have 8° angles. To produce thicker wedges, loosen the wing nut and slide the stop block slightly away from the kerf. Tighten the wing nut and cut the wedges *(inset)*.

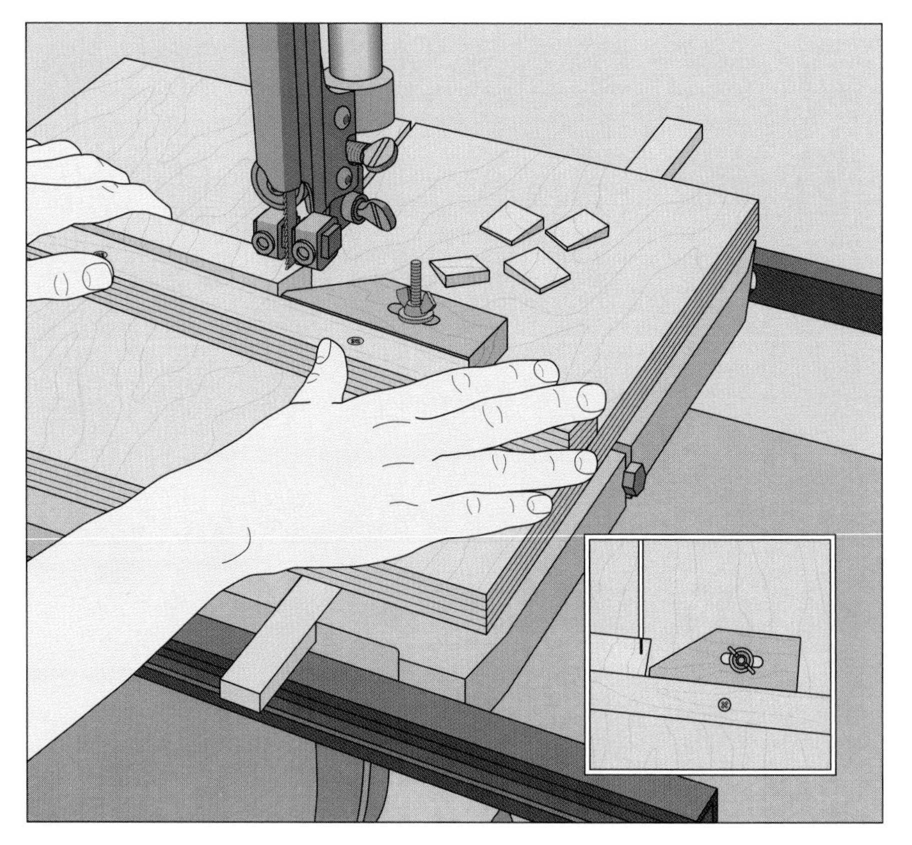

TWO JIGS FOR ANGLE CUTS ON THE BAND SAW

TAPER JIG

Making taper cuts

The simple L-shaped jig shown at right will enable you to cut tapers on the band saw. Mark the desired taper on the workpiece and place it on a board with a perfectly square edge, aligning the marked line with the board's edge. Use the long edge and the end of the workpiece as a straightedge to mark an angled cutting line and the lip on the board. Saw along the cutting line, stopping 2 inches from the end of the cut at the bottom end of the board. Turn the board 90° to cut out the lip. To use the board as a jig, set up the band saw's rip fence to the right of the blade and hold the jig flush against the fence. Align the edge of the jig's lip with the saw blade and lock the fence in position. Seat the workpiece against the jig. Turn on the saw and slide the workpiece and the jig together across the table into the blade *(right)*, keeping both hands clear of the cutting edge.

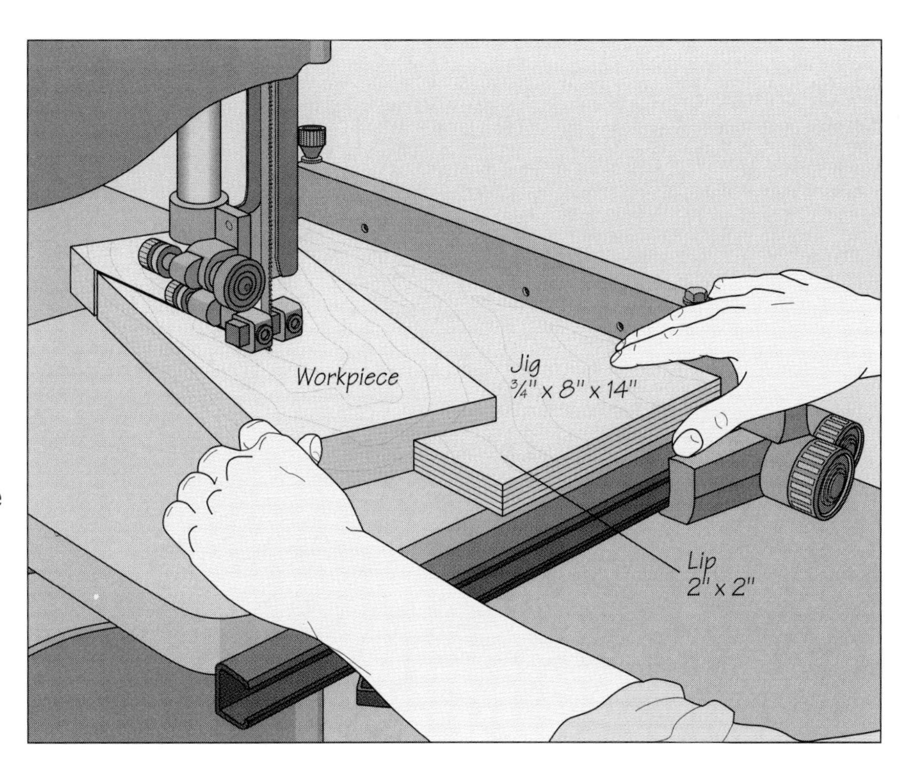

MITER JIG

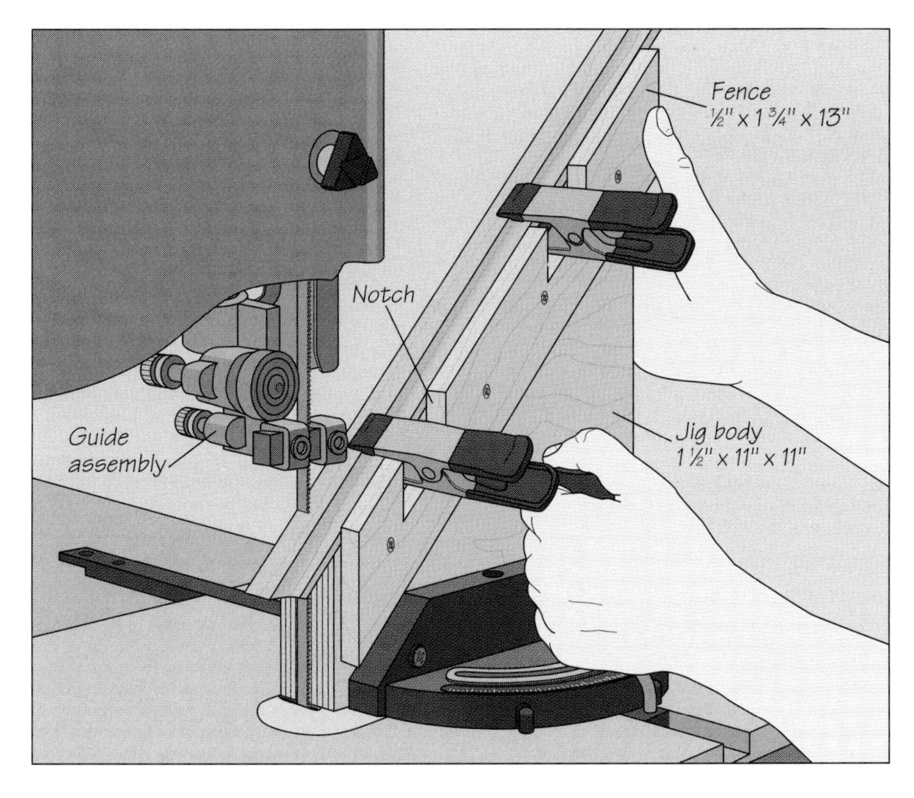

Mitering trim

Use the jig shown at left to miter trim on the band saw without angling the miter gauge. Form the jig body by face-gluing two square pieces of ¾-inch plywood together. Once the adhesive has dried, cut a 45° miter from corner to corner across the body, forming a ledge on which the workpiece will sit. Cut two slots into the face of the body ¼ inch below the angled ledge to accommodate spring clamp jaws. Next, cut the fence from solid stock, notch it for the clamps, and attach it to the jig body so that its top edge extends ¼ inch above the ledge. Screw the jig to the miter gauge and feed the jig into the blade to trim the lower end. To cut a miter, clamp the workpiece face-down on the ledge and flush against the fence, and feed the jig foward with the miter gauge *(left)*. Be sure to raise the saw's guide assembly high enough to avoid hitting the jig or workpiece.

A taper jig for the radial arm saw can be built exactly like the table saw jig shown below, with one addition: Because the saw motor is suspended above the table, the radial arm saw jig can only be fed with one hand, making a handle necessary.

A TAPER JIG FOR THE TABLE SAW

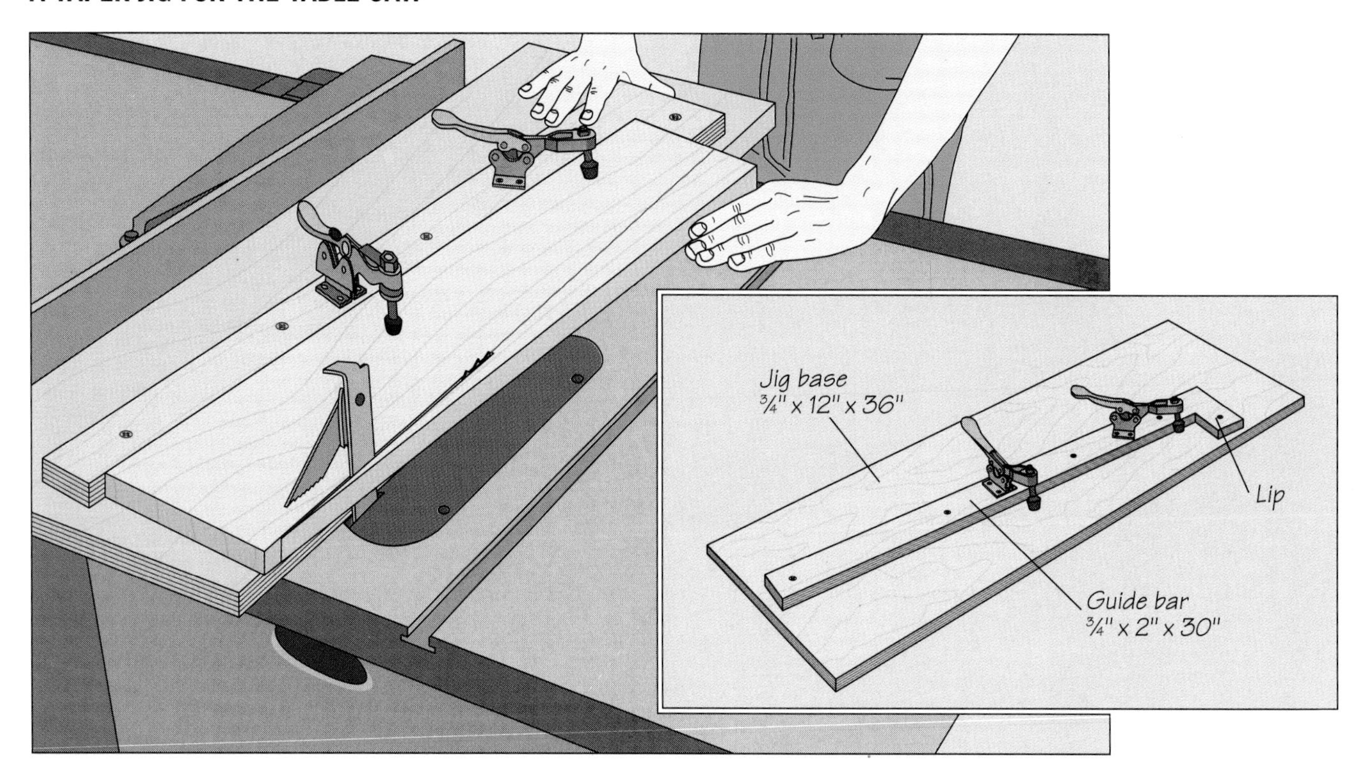

1 Making a taper cut
For accurate taper cuts on the table saw, build this jig *(inset)* from ¾-inch plywood. Refer to the illustration for suggested dimensions. To assemble the jig, set the saw blade to its maximum cutting height, butt one side of the jig base against the blade and position the rip fence flush against the other side of the base. Lower the blade and mark a cutting line for the taper on the workpiece, then set it on the base, aligning the line with the edge of the jig base nearest the blade. Holding the workpiece securely, position the guide bar against it, with the lip snugly against the end of the workpiece. Screw the guide bar to the base and press the toggle clamps down to secure the workpiece to the jig. To make the cut, set the blade height and slide the jig and workpiece across the table, making sure that neither hand is in line with the blade *(above)*. **(Caution: Blade guard removed for clarity.)**

A BLADE HEIGHT GAUGE

SETTING THE BLADE HEIGHT ON A TABLE SAW

Using a blade height gauge
Your table saw's blade can be set at a specific height quickly with a blade height gauge. Make the jig from strips of 1/8- or 1/16-inch-thick hardboard or solid wood laminated together. First, rip a length of the stock to a width of 3 inches. Crosscut the piece into strips, starting with an 8-inch length. Make each successive strip 3/8 inch shorter than the previous one. Once all the strips are cut, glue them together face-to-face with one end aligned. To use the jig, set it on the saw table beside the blade and rotate the blade height adjustment crank until the blade contacts the gauge at the desired height *(right).*

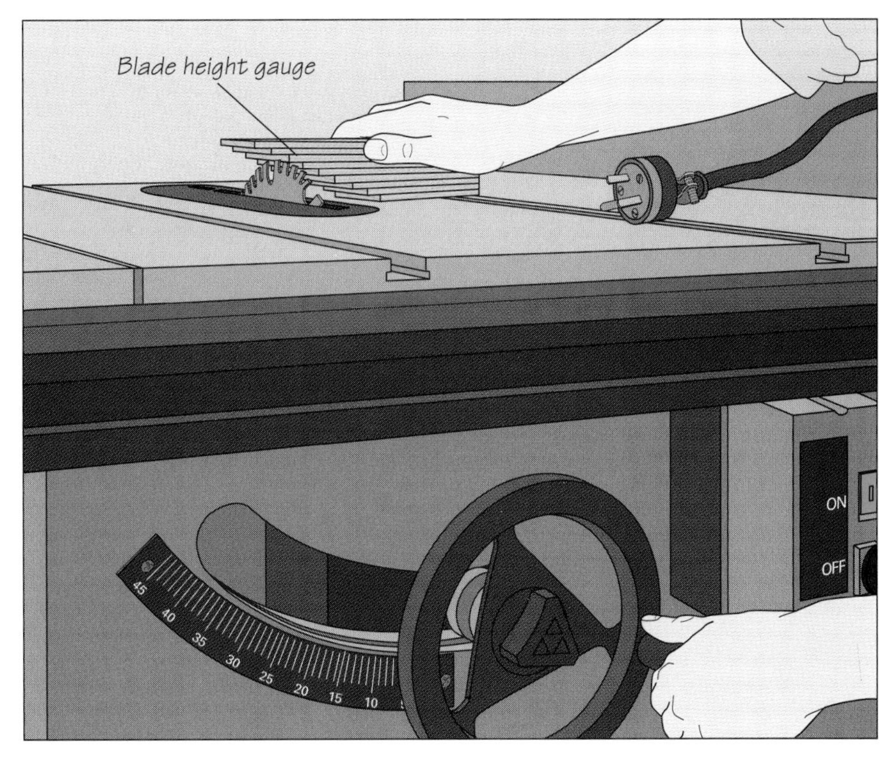

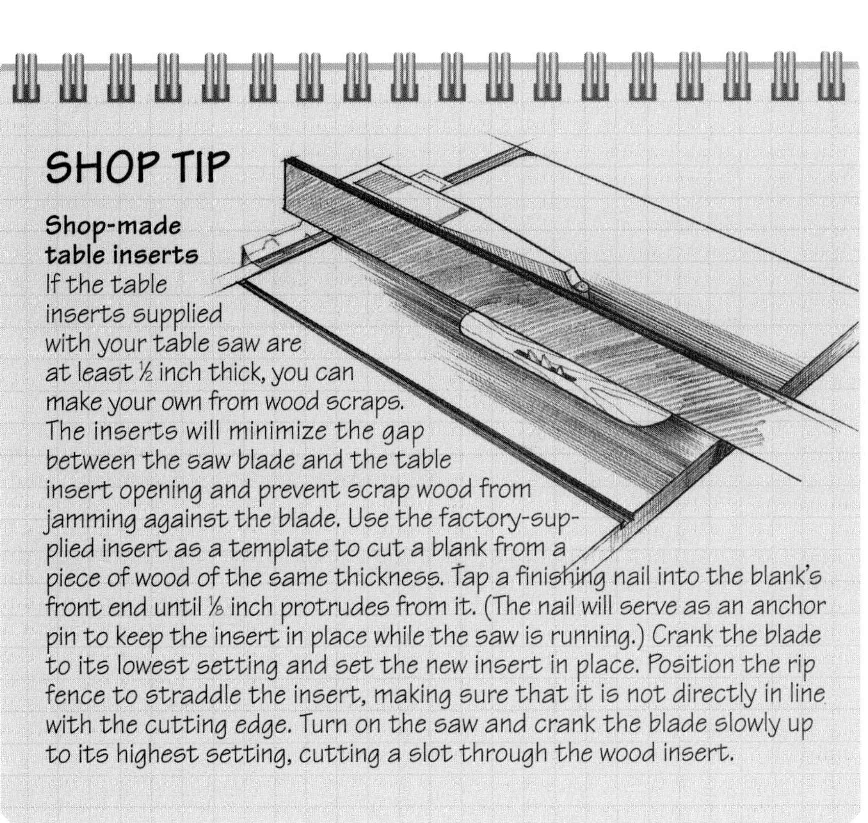

SHOP TIP

Shop-made table inserts
If the table inserts supplied with your table saw are at least 1/2 inch thick, you can make your own from wood scraps. The inserts will minimize the gap between the saw blade and the table insert opening and prevent scrap wood from jamming against the blade. Use the factory-supplied insert as a template to cut a blank from a piece of wood of the same thickness. Tap a finishing nail into the blank's front end until 1/8 inch protrudes from it. (The nail will serve as an anchor pin to keep the insert in place while the saw is running.) Crank the blade to its lowest setting and set the new insert in place. Position the rip fence to straddle the insert, making sure that it is not directly in line with the cutting edge. Turn on the saw and crank the blade slowly up to its highest setting, cutting a slot through the wood insert.

The table saw miter jig shown at right is similar to the crosscut jig described below, except that instead of an extension and safety block it features two 12-inch-long 1-by-4 miter arms. Placed at 90° to each other in the middle of the jig, the arms ensure that a workpiece mitered along one guide will form a perfect 90° corner with a board cut along the other arm.

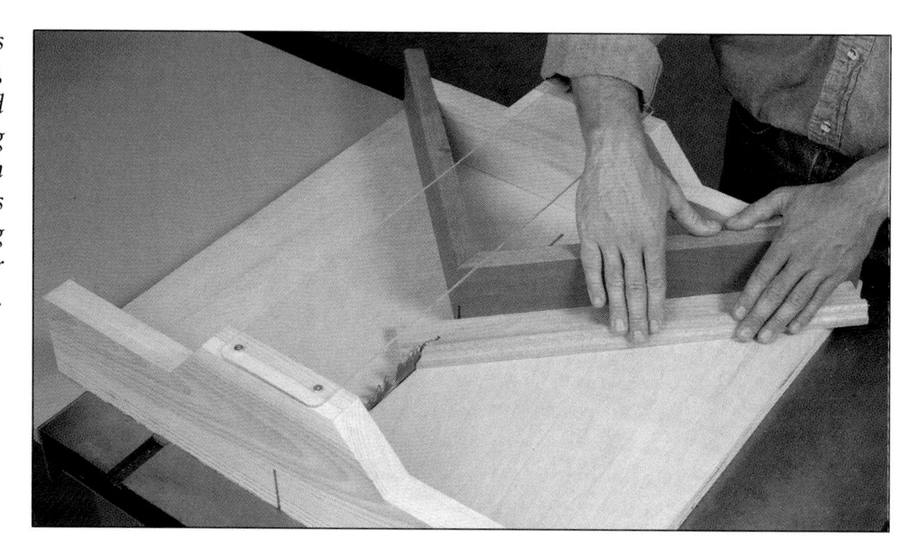

A CROSSCUT JIG FOR THE TABLE SAW

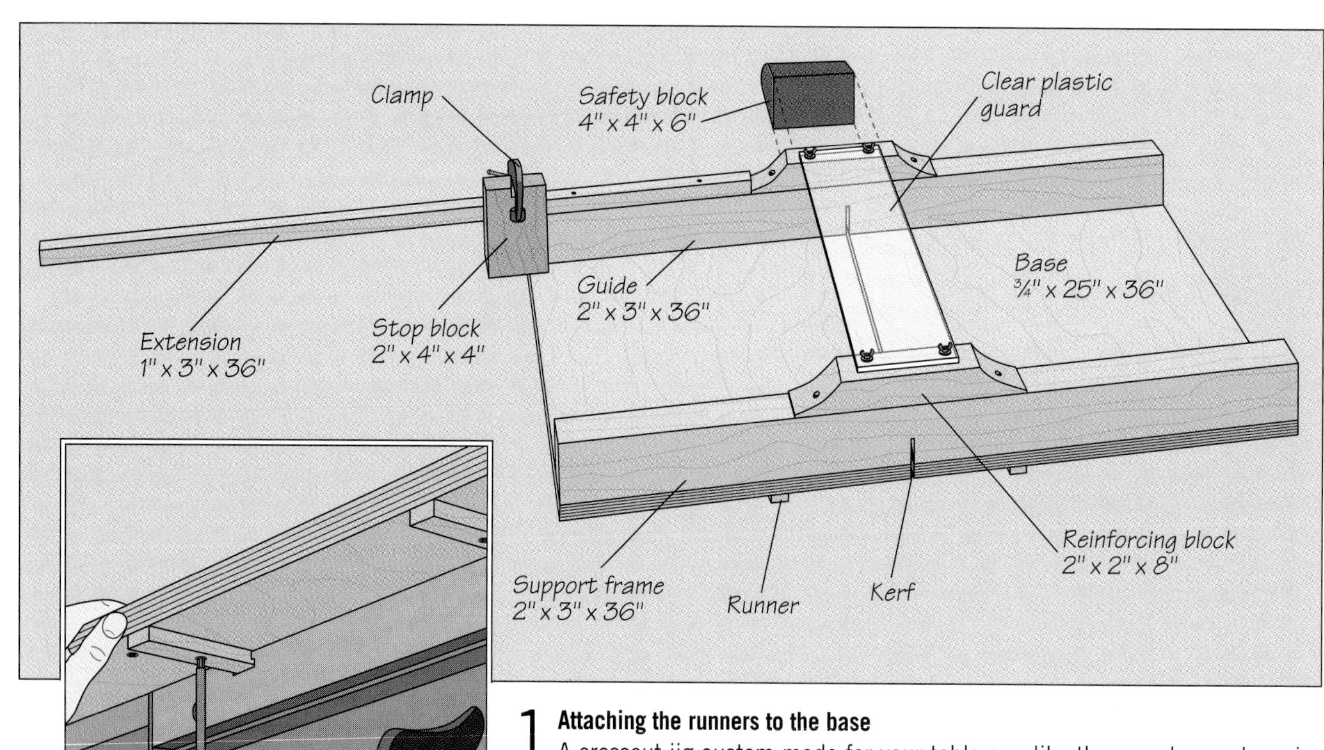

1 Attaching the runners to the base
A crosscut jig custom-made for your table saw like the one shown above is especially valuable if you are working with unwieldy stock. Refer to the illustration for suggested dimensions. Start by cutting two 25-inch-long hardwood runners to fit your miter slots. Bore and countersink clearance holes for screws into the undersides of the runners, 3 inches from each end. Place the runners in the slots and slide them out to overhang the back end of the table by about 8 inches. Position the jig base squarely on the wood strips, its edge flush with their overhanging ends, and screw the runners to the base *(left)*. Slide the runners and the base off the front end of the table and drive in the other two screws.

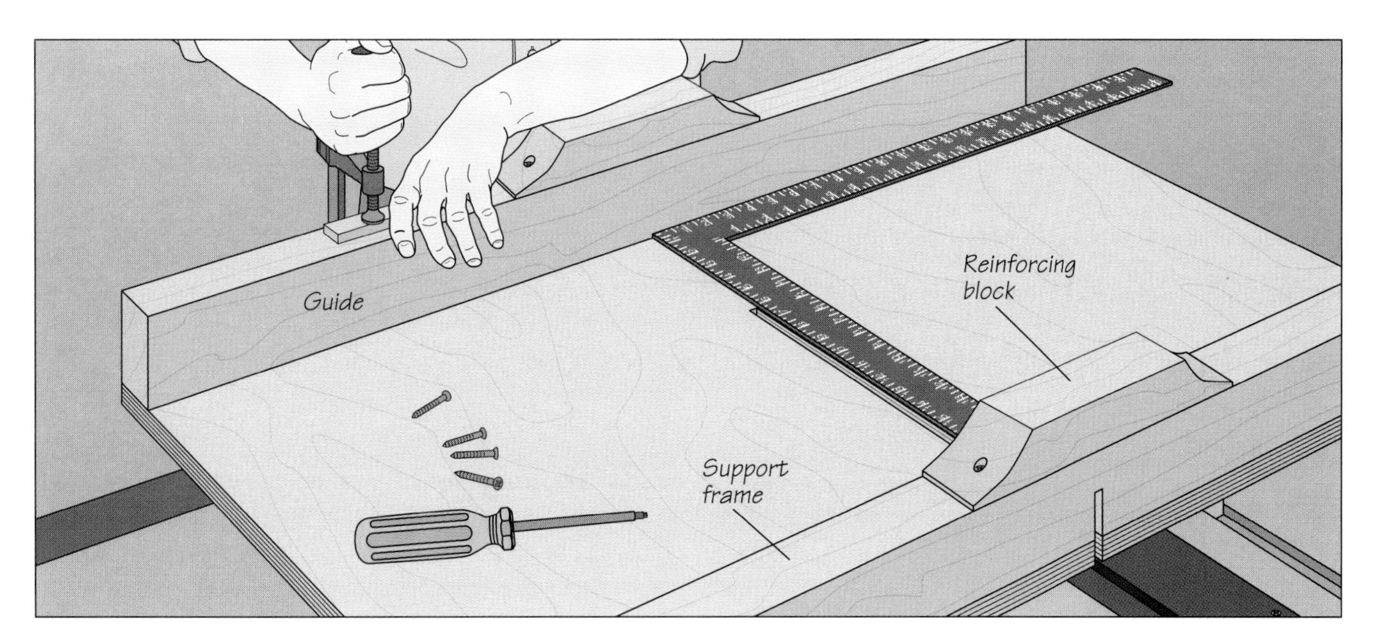

2 Installing the support frame and guide
With the runners still in the miter slots, attach the support frame along the back edge of the jig and glue on the reinforcing block, centered between the runners. Then make a cut through the support frame and three-quarters of the way across the base. Turn off the saw and lower the blade. Screw a reinforcing block to the guide and position the guide along the front edge of the jig, using a carpenter's square to ensure that it is square with the saw kerf. Clamp the guide in place *(above)* and screw it to the base from underneath the jig, making sure you countersink the fasteners. Glue the safety block to the outside face of the guide, again centered on the kerf. Raise the saw blade and finish the cut, sawing completely through the guide but only slightly into the safety block. Mount a clear plastic sheet over the saw kerf as a blade guard, fastening it to the reinforcing blocks with wing nuts or screws.

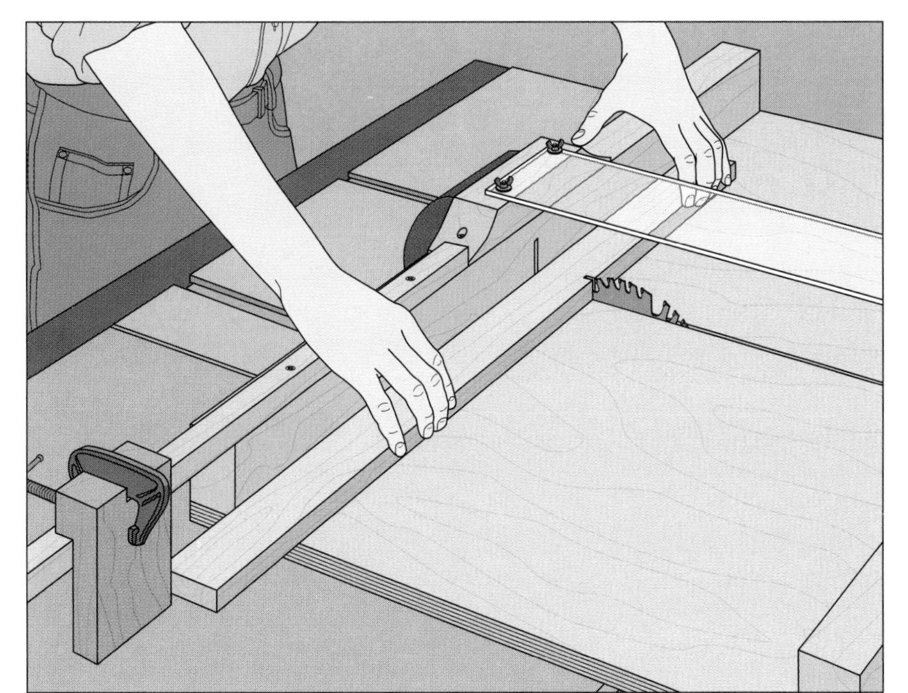

3 Crosscutting
For making repeat cuts to the same length, screw an extension to the right side of the guide and clamp a stop block to it. Cut a notch in the block to hold the clamp in place when it is loosened. To use the jig, fit the runners into the miter slots and slide the jig toward the back of the table until the blade enters the kerf. Hold the workpiece against the guide, slide the stop block to the desired position, and clamp it in place. With the workpiece held firmly against the guide and the stop block, slide the jig steadily across the table *(left)*, feeding the workpiece into the blade.

RAISED PANEL JIG

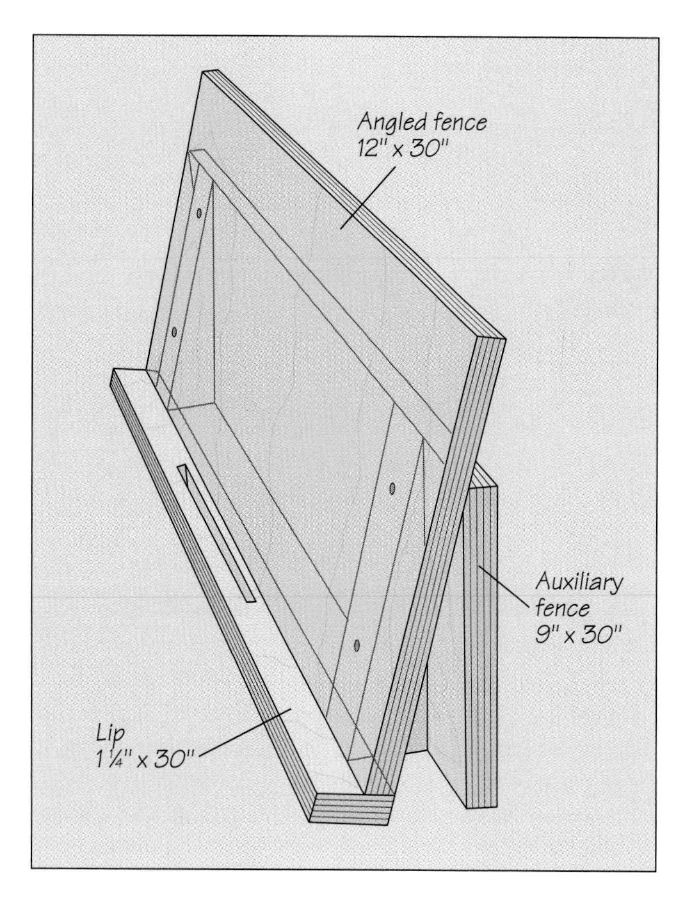

1 Making the jig

To raise a panel on the table saw without adjusting the blade angle, use the shop-built jig shown at left. Refer to the illustration for suggested dimensions. Screw the lip along the bottom edge of the angled fence, making sure to position the screws where they will not interfere with the blade. Prop the angled fence against the auxiliary fence at the same angle as the cutting line marked out on the panel to be raised. Use a sliding bevel to transfer this angle to triangular-shaped supports that will fit between the two fences and cut the supports to fit. Fix the supports in place with screws *(above)*.

2 Raising a panel

Shift the rip fence to position the jig on the saw table with the joint of the lip and angled fence over the blade; ensure that the screws are well clear of the table opening. Turn on the saw and crank the blade up slowly to cut a kerf through the lip. Next, seat the panel in the jig and adjust the height of the blade until a single tooth is protruding beyond the front of the panel. Make a test cut in a scrap board the same thickness as the panel and then test its fit in the frame groove. Adjust the position of the fence or blade, if necessary, and cut the actual panel, beveling the end grain first *(right)*.

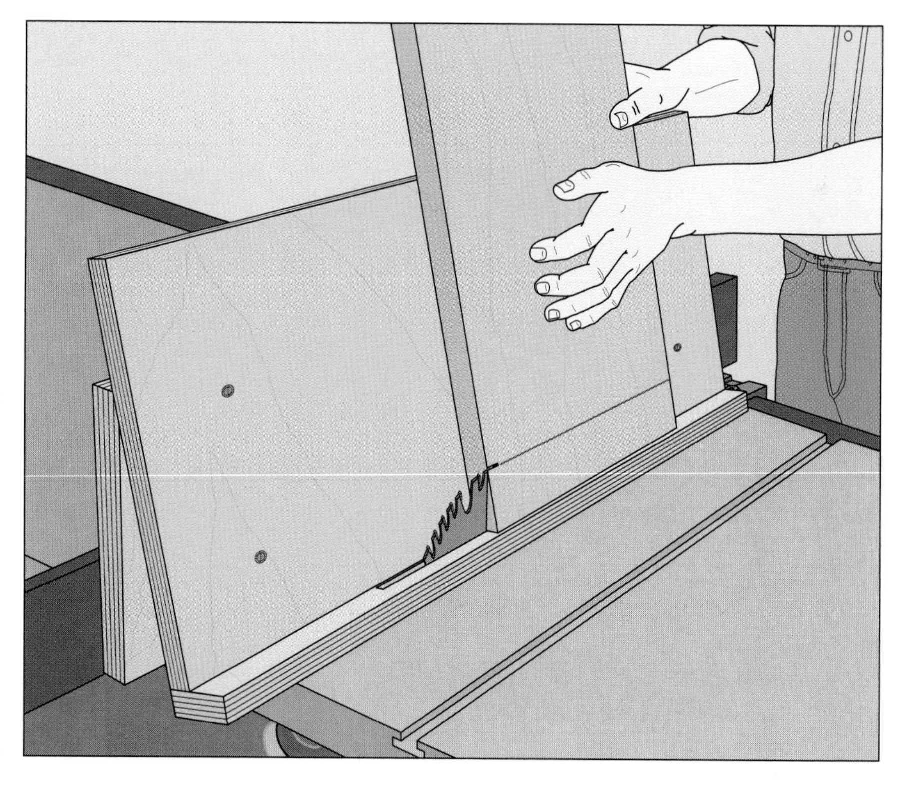

BOARD-STRAIGHTENING JIG

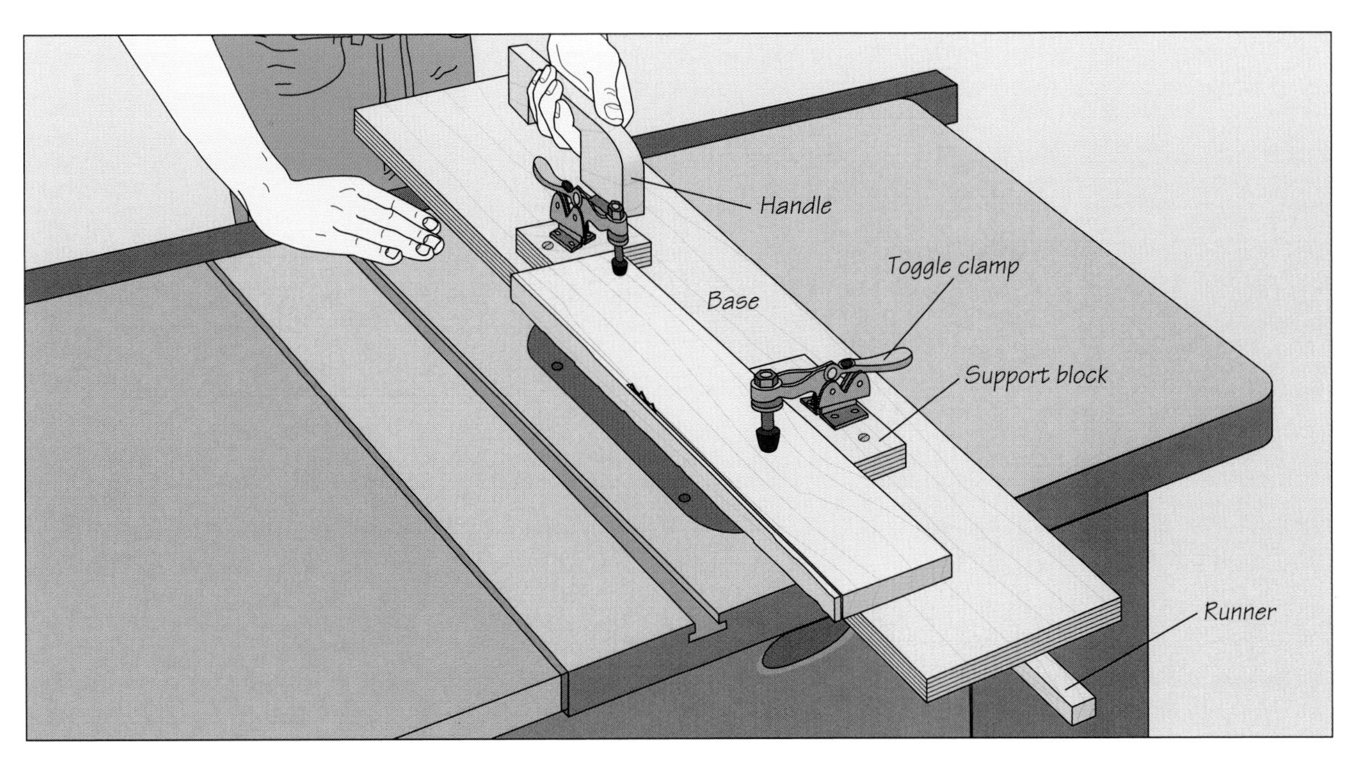

Truing a board

To true uneven boards on a table saw, build the board straightening jig shown above. Built from ¾-inch plywood, the jig slides in the table saw's miter gauge slot, while the board to be straightened is held in place by support blocks and toggle clamps. First cut the base from ¾-inch plywood; make it about 9 inches wide and longer than the width of your saw table. Cut a runner to fit the left-hand miter gauge slot; make it longer than the jig base and position it on the underside of the base so that the inside edge of the base overlaps the blade by ⅛ inch. Screw the runner to the bottom of the jig, countersinking all the screws. Next, screw two support blocks to the base and install toggle clamps on them; position the support blocks so the workpiece is centered on the base. Finally, fashion a handle and attach it to the end of the jig. To use the jig, first trim the inside edge square by running it across the blade, then clamp the board to be straightened to the jig and repeat to true its edges *(above)*.

SHOP TIP

Crosscutting wide panels

If your table saw does not have an extension, or you are working with panels that are too large to cut with a miter gauge, you can crosscut these panels by clamping a square 1-by-3 fence to the underside of the panel. Use a 1-by-3 that is longer than the panel is wide and carefully position it underneath perpendicular to the panel's long edge, then clamp the fence to the panel with C clamps. Guide the fence along the outer edge of your saw's table as you make the cut.

A RADIAL ARM SAW MOLDING FIXTURE

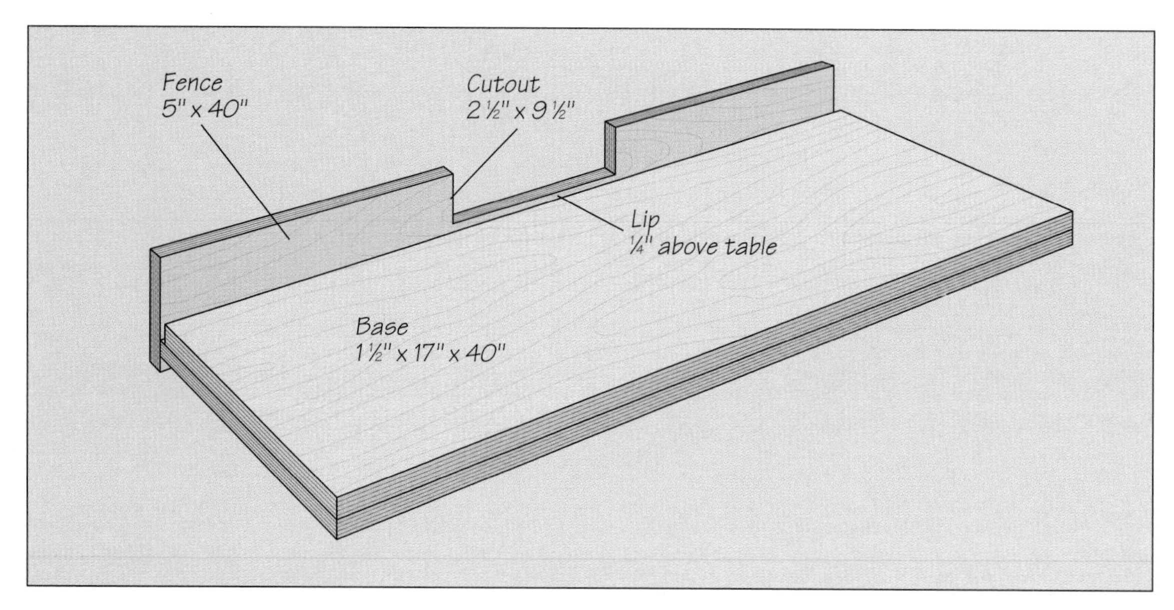

1 Building and setting up the jig

If you want to cut grooves or molding on the radial arm saw with the blade in the horizontal position, try an auxiliary fence and table like the one shown above. The base raises the workpiece to the cutter, and the fence supports the work while providing a clearance cutout for the dado or molding head guard. Cut the fence and the two pieces for the base from ¾-inch plywood; make the base pieces the same size as the front saw table. Screw the base pieces together, offsetting the top slightly to create a gap along the fence that will prevent sawdust from accumulating between the base and the fence. When sawing the fence cutout, leave a lip that will protrude at least ¼ inch above the base when the jig is installed. The lip will support the workpiece as it rides along the fence during a cut. Screw the fence to the base. To set up the jig, slip the fence between the front saw table and the table spacer, then tighten the table clamps to secure it.

2 Cutting a molding

Install a molding head on your saw, then secure the workpiece by clamping one featherboard to the outfeed side of the fence and another to the table, braced with a support board. Adjust the molding head for a ⅛-inch-deep cut, making certain the blade guard is positioned just above the workpiece. Feed the stock into the cutters with your right hand *(right)*; use your left hand to press the workpiece against the fence. Finish the pass with a push stick. Make as many passes as necessary, advancing the molding head no more than ⅛ inch into the workpiece at a time. Once you have cut the desired profile, make a final, very shallow pass, feeding slowly and evenly to produce a smooth finish.

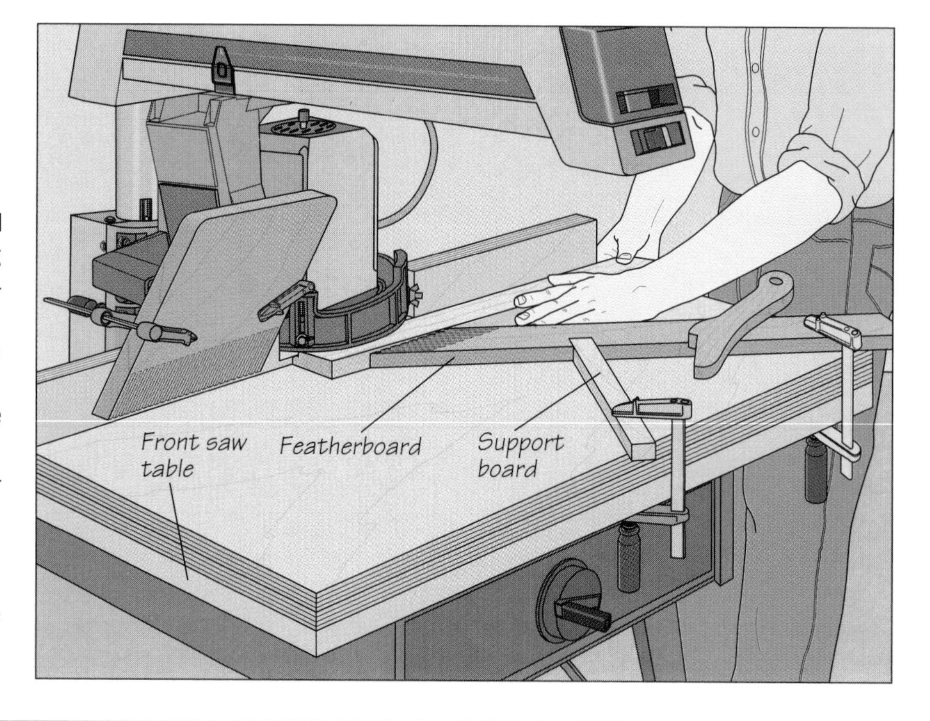

AN AUXILIARY FENCE FOR THE TABLE SAW

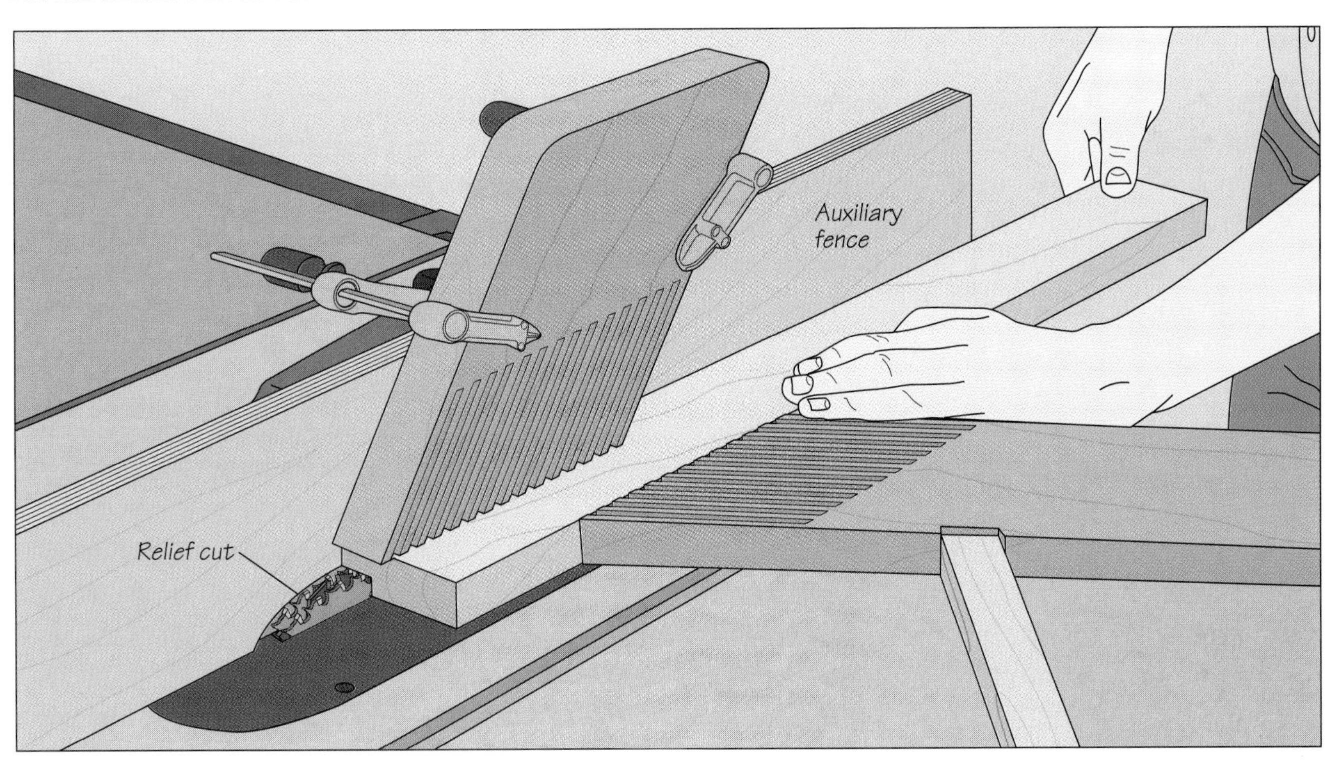

Using an auxiliary fence
Anytime you use a dado or molding head on your table saw in conjunction with the rip fence, you need to attach a wood fence to prevent the blades from contacting the metal one. Cut the fence from ¾-inch plywood the same length as the rip fence and slightly higher. Lower the blades below the saw table and screw the wood fence to the rip fence. Mark the depth of cut on the fence and position the auxiliary fence directly over the dado head, ensuring that the metal fence is clear of the blade. Turn on the saw and slowly crank up the dado head to the marked line, producing a relief cut in the auxiliary fence. Butt the cutting line on the workpiece against the outer blade and lock the fence flush against the stock. For most cutting operations, like the rabbet cut shown in the illustration, the workpiece should be supported with featherboards. Feed the workpiece at a steady rate, using both hands *(above)*.

SHOP TIP

A miter gauge extension
To increase the bearing surface of your table saw's miter gauge or to make repeat crosscuts with the aid of a stop block, attach a wood extension to the gauge. The extension must be long enough to extend beyond the saw blade. Screw it to the gauge and push the gauge to cut off the end of the extension. Turn off the saw and slide the miter gauge to the front of the table. To set up a stop block, clamp a wood block to the extension the desired distance from the blade. For each cut, butt the edge of the workpiece against the extension and the end against the stop block.

COVE-CUTTING JIG

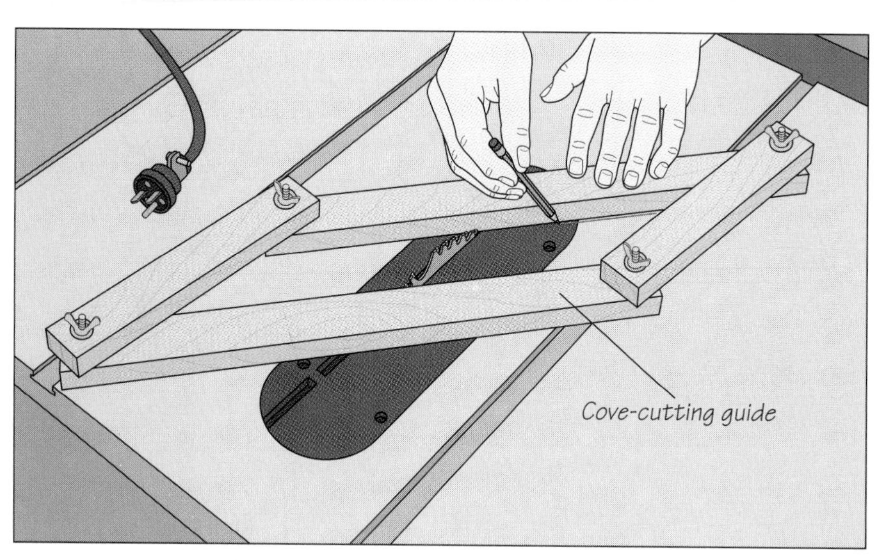

1 Building and using the jig
Fashion molding on the table saw with the help of the cove-cutting guide shown at left. To construct the jig, fasten two 18-inch-long 1-by-2s to two 9-inch-long 1-by-2s with carriage bolts and wing nuts, forming two sets of parallel arms. Adjust the jig so the distance between the inside edges of the two long arms equals the width of the cove. Crank the blade to the desired depth of cut. Lay the guide across the blade and rotate it until the blade, turned by hand, just touches the inside edges of the arms. Then run a pencil along the inside edges of the long arms to trace guidelines across the table insert *(left)*.

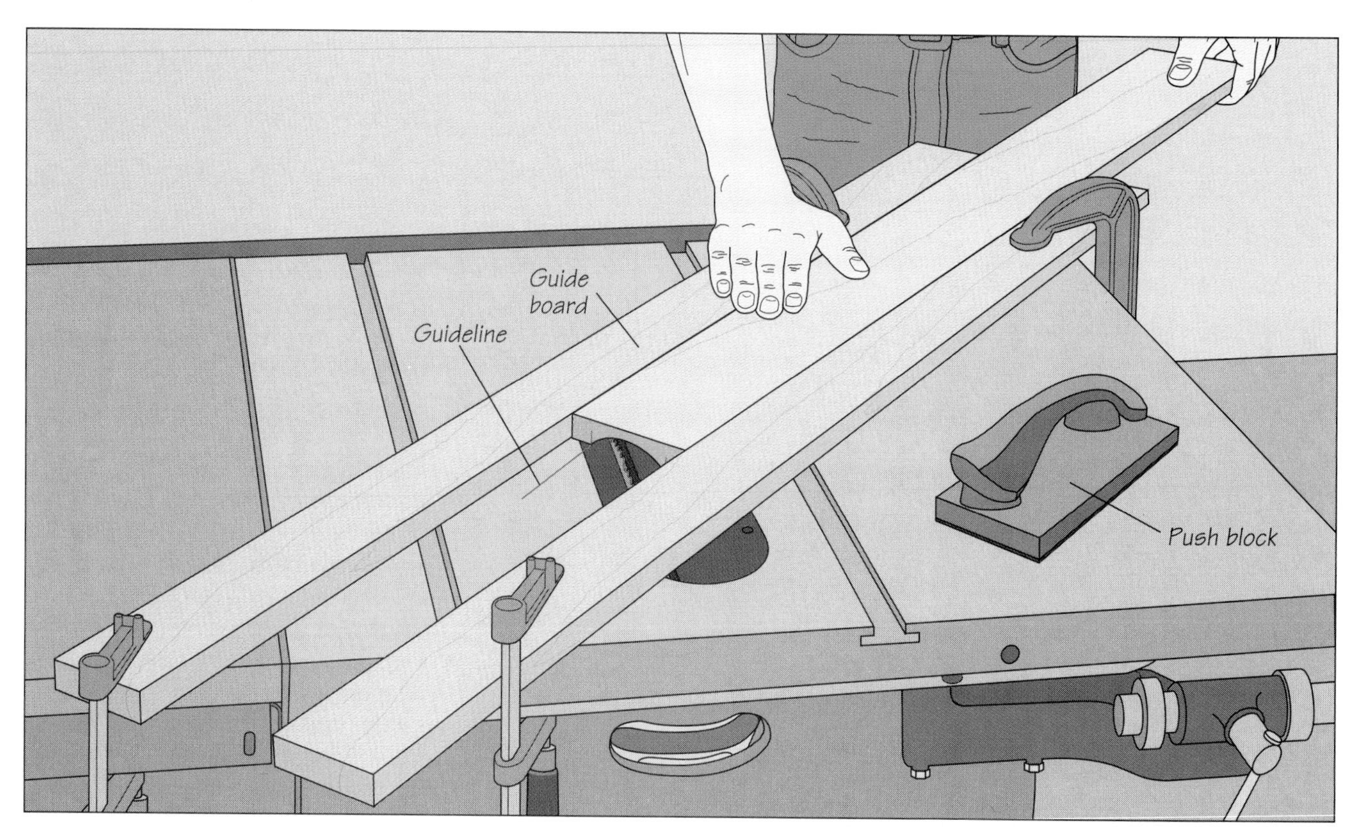

2 Cutting a cove
Remove the guide and lower the blade beneath the table. Outline the desired cove profile on the leading end of the workpiece, then set the stock on the saw table, aligning the marked outline with the guidelines on the table insert. Butt guide boards against the edges of the workpiece and clamp them parallel to the guidelines; use boards long enough to span the saw table. Crank the blade ⅛ inch above the table. To make the first pass, feed the workpiece steadily *(above)*, using push blocks when your hands approach the blade area. Make as many passes as necessary, raising the blade ⅛ inch at a time.

A SIMPLE TENONING JIG

1 Building the jig

Easy to assemble, the fence-straddling jig shown at right works well for cutting two-shouldered open mortise-and-tenon joints. Refer to the dimensions suggested in the illustration, making sure the thickness of the spacer and width of the brace allow the jig to slide smoothly along your rip fence without wobbling. Cut the body and brace from ¾-inch plywood and the guide and spacer from solid wood. Saw an oval hole for a handle in one corner of the jig body and attach the guide to the body directly in front of the handle hole, making sure that the guide is perfectly vertical. (The blade may notch the bottom of the guide the first time you use the jig.) Screw a small wood block to the body below the handle and attach a toggle clamp to the block. Finally, fasten the spacer and brace in place.

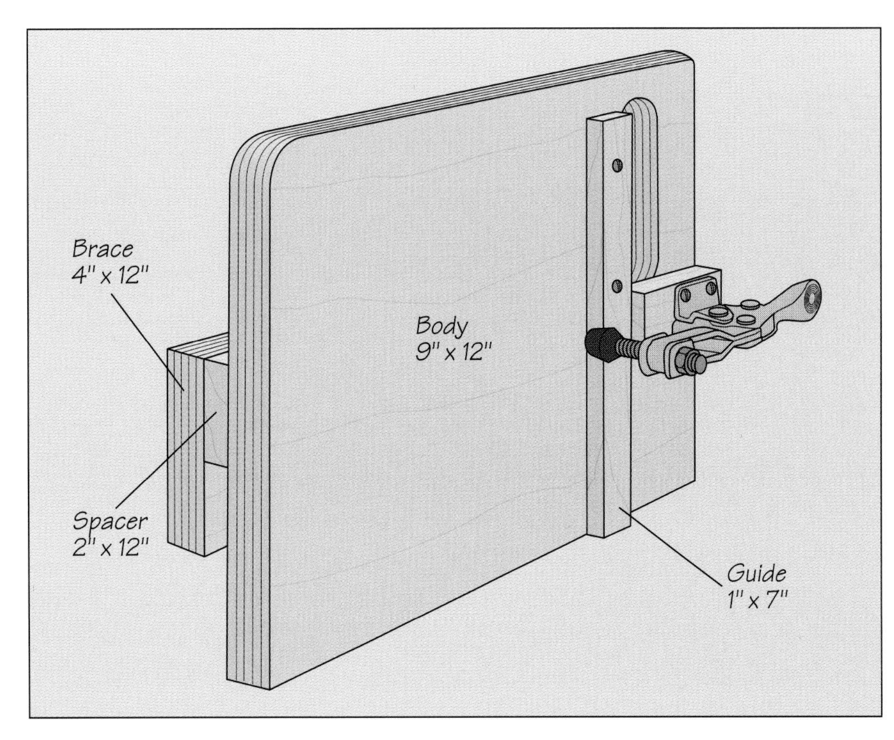

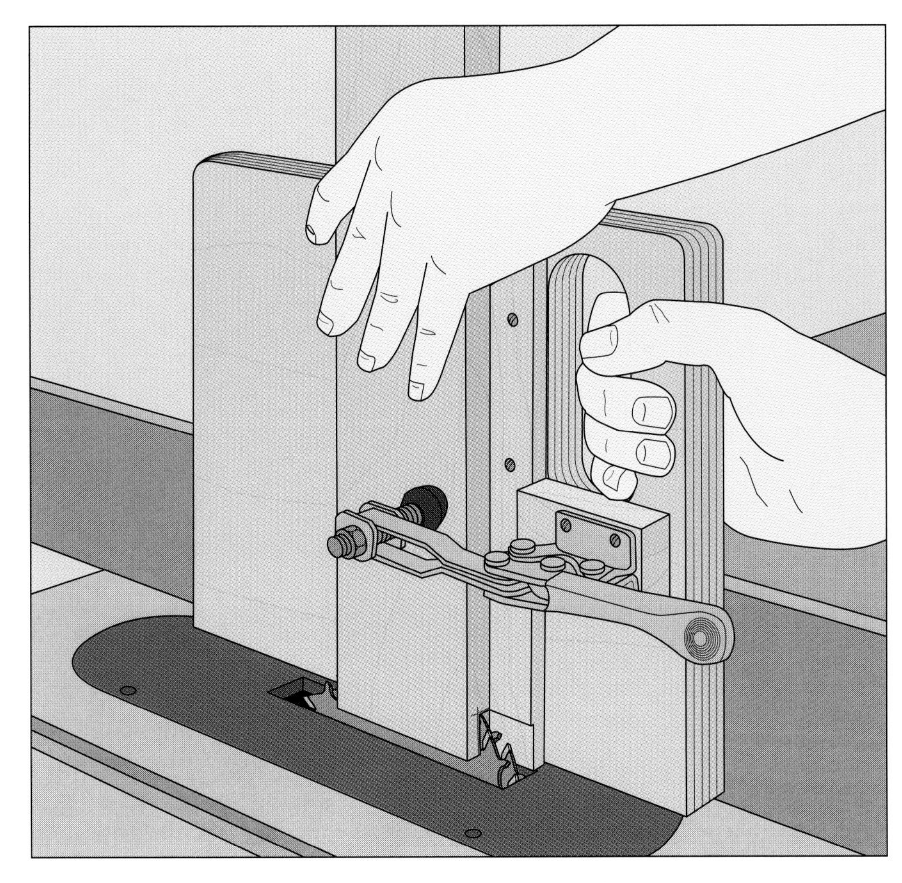

2 Cutting a mortise

Place the jig astride the fence. Butt the workpiece against the jig guide and clamp it in place. Position the fence to align the cutting marks on the board with the blade and slide the jig along the fence to make the cut *(left)*.

AN ADJUSTABLE TENONING JIG

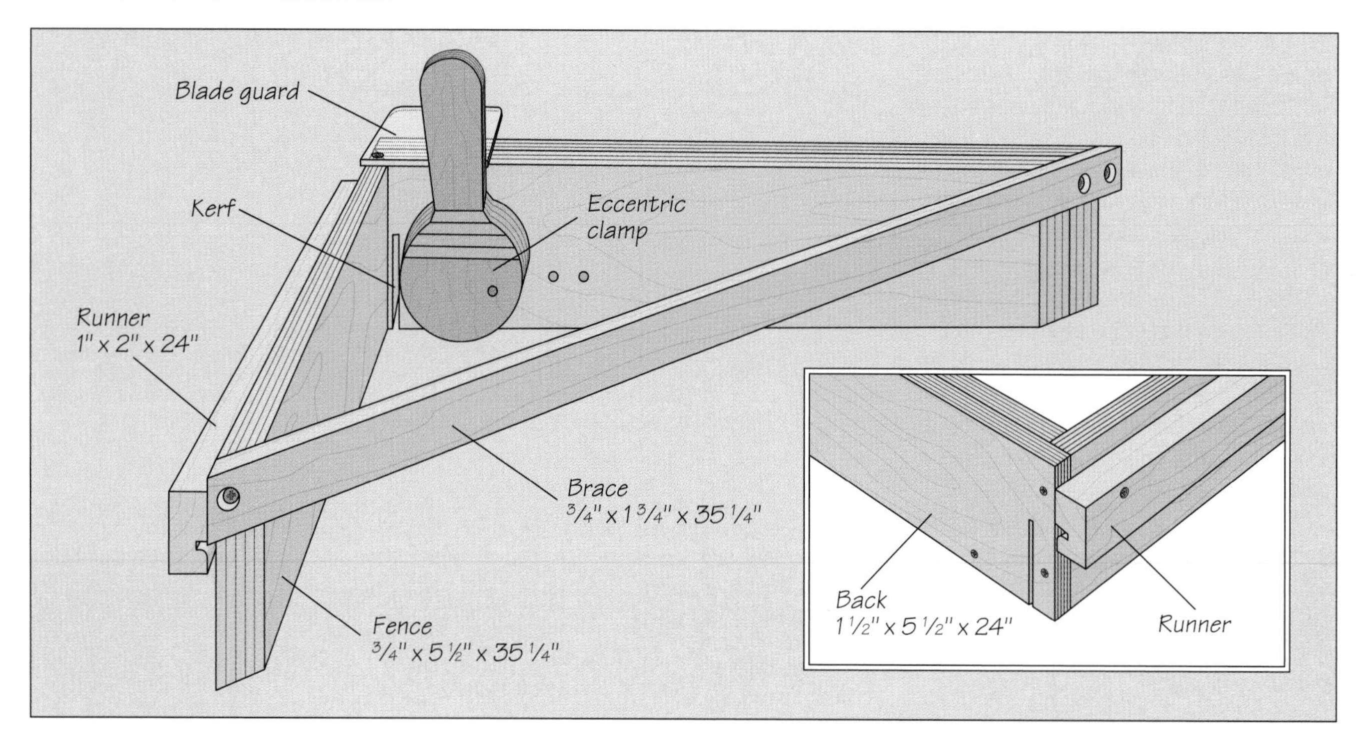

1 Assembling the jig
The jig shown above can be used on the table saw to cut both parts of an open mortise-and-tenon joint. Refer to the illustration for suggested dimensions. Cut the jig fence and back from three pieces of ¾-inch plywood and saw a 45° bevel at one end of each board; the pieces should be wider than the height of your saw's rip fence. Fasten two pieces together face-to-face to fashion the back, then use countersunk screws to attach the fence and back together in an L shape; make sure the fasteners will not be in the blade's path when you use the jig *(inset)*. Next, cut the brace from solid stock, bevel its ends and attach it along the top edges of the fence and back, forming a triangle. Cut the runner from solid wood and attach it to the fence so that the jig runs smoothly across the table without wobbling. (The runner in this illustration has been notched to fit the particular design of the saw's rip fence.) Finally, cut a piece of clear plastic as a blade guard and screw it to the jig back flush with its front face.

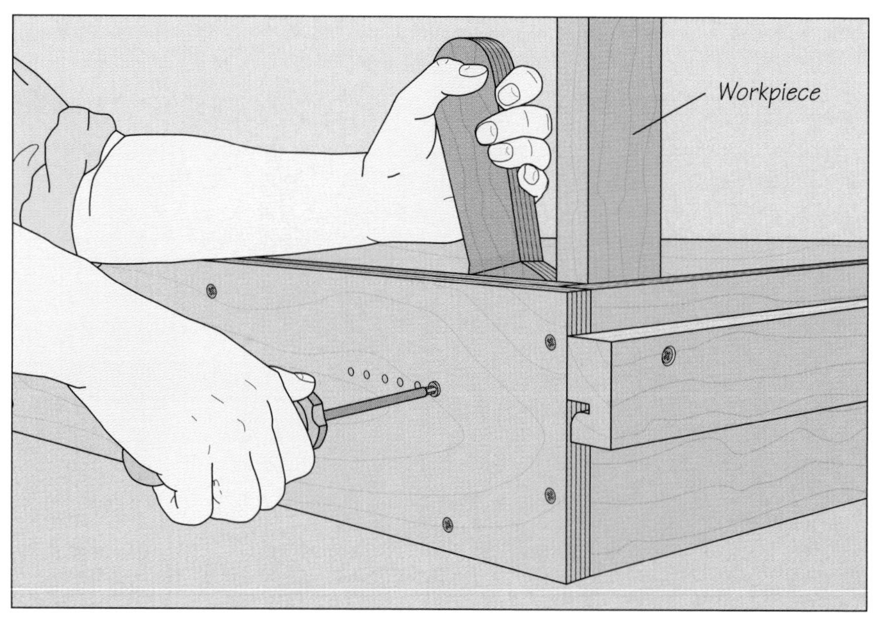

2 Mounting the eccentric clamp
Make the clamp by face-gluing three pieces of ¾-inch plywood and cutting the assembly into the shape shown. Bore a pilot hole through the jig back and the clamp, then fasten the clamp in place; wedge one of your workpieces between the edge of the clamp and the fence as you drive the screw. Offset the fastener so the clamp can pivot eccentrically *(above)*. (Drill additional holes in the jig back to enable you to move the clamp to accommodate stock of varying thicknesses.)

3 Cutting a tenon
Set the jig on the saw table in front of the blade with the runner and fence straddling the rip fence. Secure the workpiece in the jig by turning the eccentric clamp, and position the rip fence so that the blade is in line with a tenon cheek cutting mark on the workpiece. Feed the jig into the blade. (Your first use of the jig will produce a kerf in the back.) Flip the workpiece in the jig and repeat to cut the other cheek *(above)*. Remove the jig from the table, lower the cutting height to the level of the shoulders, and shift the rip fence to cut the tenon shoulders.

SHOP TIP

A miter gauge angle-setting jig
To keep track of non-standard angles that you commonly use on your table saw's miter gauge, make a set of angle-setting jigs. Simply cut two 1-by-2s and clamp them to the miter gauge, one against the bar and one against the face. Screw them together into an angled L shape and mark down the angle they form on the jig. Use the device like a sliding bevel to set the miter gauge quickly to a specific angle.

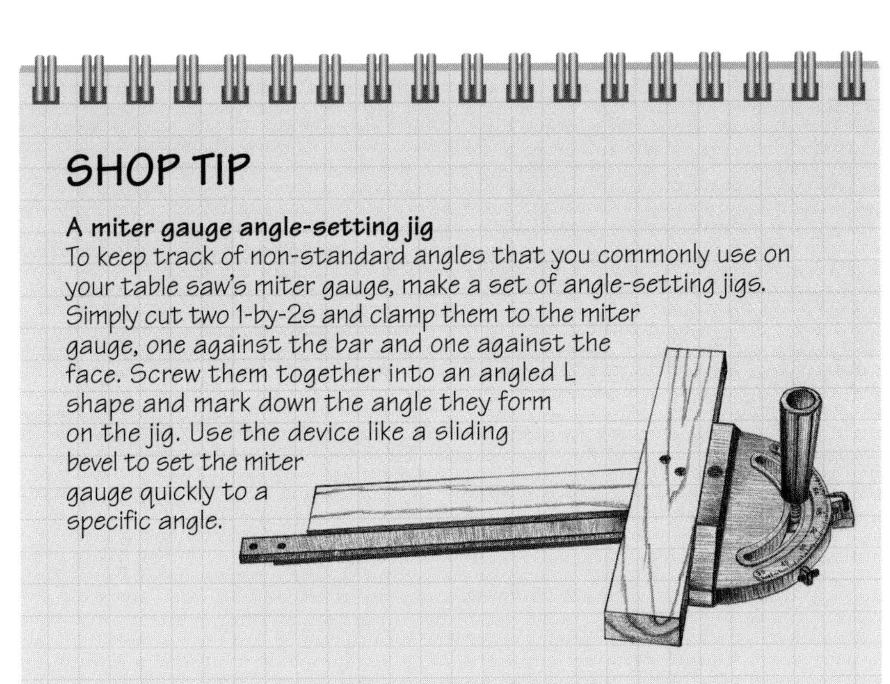

DRILLING JIGS

Drilling a hole is simple, but there are times when a jig can save you time and frustration, especially when the hole must be located precisely, or at an exact angle. The center-drilling jig, for example, ensures that dowel joint holes will be centered in the edges of mating workpieces. The jig for drilling equally spaced holes on the drill press virtually eliminates the need for measuring and marking and guarantees consistently accurate results.

The tilting table and pocket hole jigs are both invaluable for boring the angled holes that are commonly used to attach tabletops and chair seats to legs and rails. The technique for drilling deep holes shown on page 67 will double your drill press's quill stroke, allowing you to bore straight through thick stock without resorting to an extension bit.

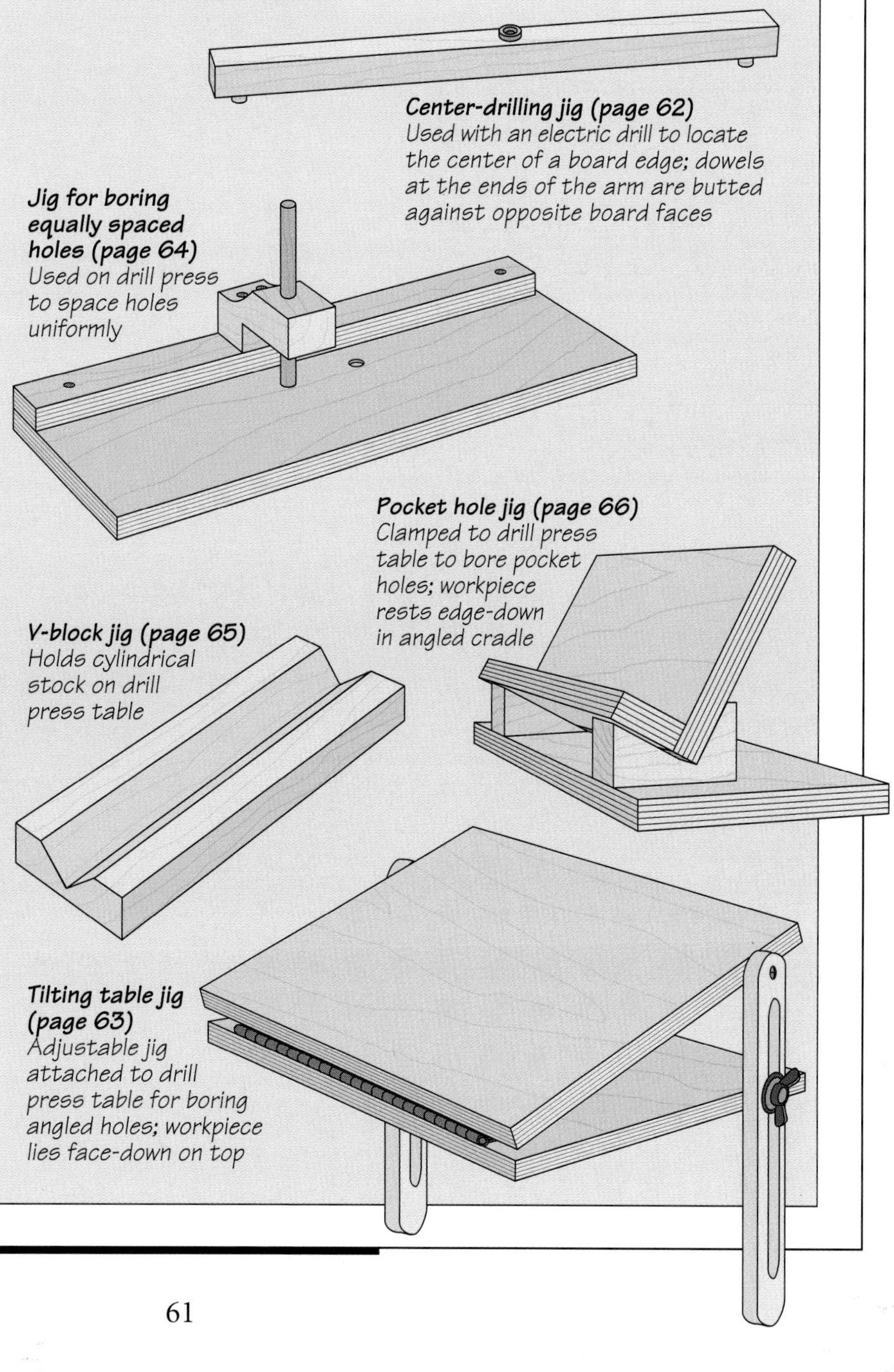

Equipped with the jig shown at left, a drill press can bore a row of equally spaced holes quickly and accurately. The jig is simple to build from scrap wood and a short dowel.

A SELECTION OF DRILLING JIGS

Center-drilling jig (page 62)
Used with an electric drill to locate the center of a board edge; dowels at the ends of the arm are butted against opposite board faces

Jig for boring equally spaced holes (page 64)
Used on drill press to space holes uniformly

Pocket hole jig (page 66)
Clamped to drill press table to bore pocket holes; workpiece rests edge-down in angled cradle

V-block jig (page 65)
Holds cylindrical stock on drill press table

Tilting table jig (page 63)
Adjustable jig attached to drill press table for boring angled holes; workpiece lies face-down on top

CENTER-DRILLING JIG

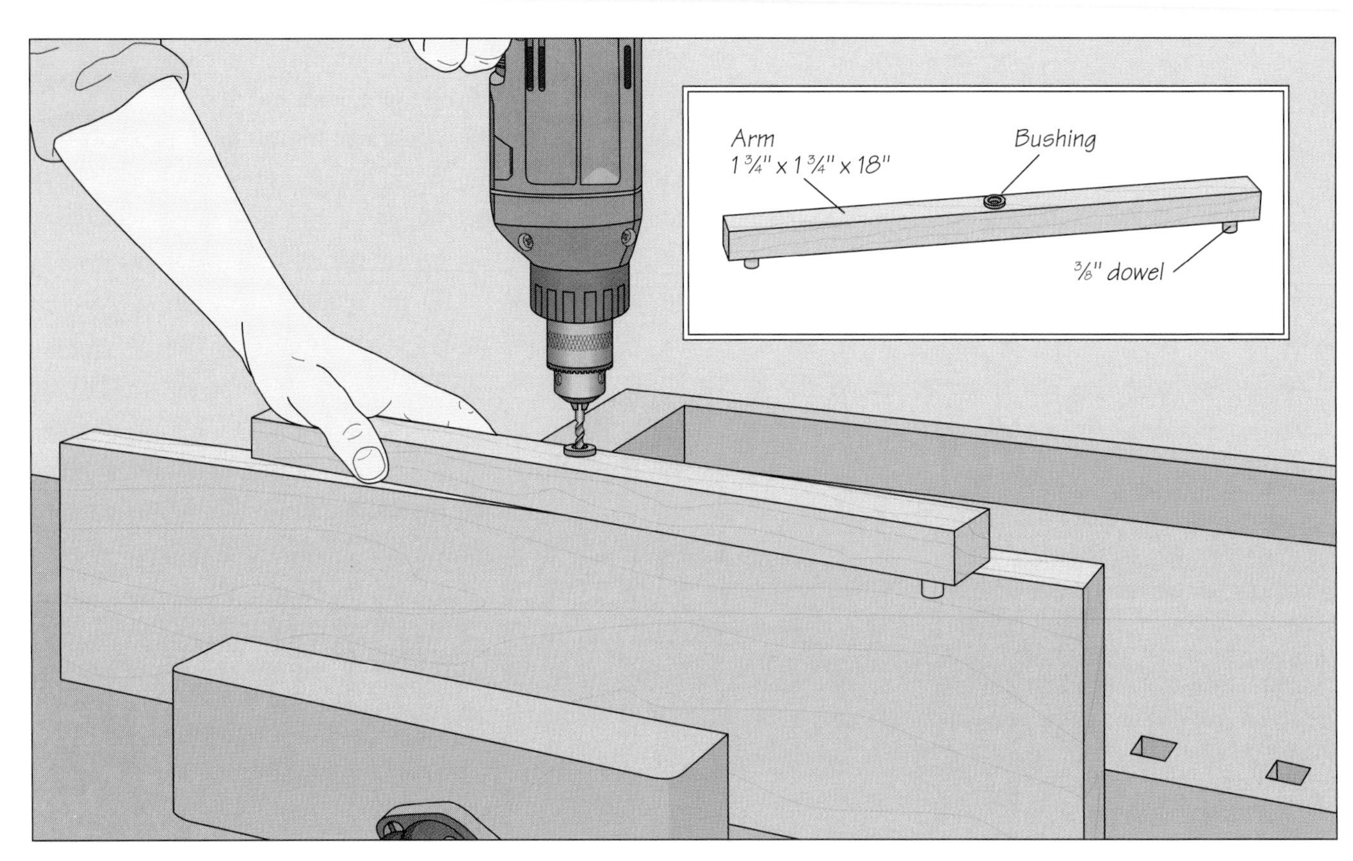

Drilling dowel holes

The simple jig shown above will enable you to bore holes that are centered on the edge of a board. The inset provides suggested dimensions. Mark the center of the top face of the arm, and bore a hole for a guide bushing. The hole in the bushing should be the same size as the holes you plan to drill. Turn the arm over and draw a line down its middle. Mark points on the line 1 inch from each end. (Check your measurements: The points must be equidistant from the center.) Then bore a 3/8-inch-diameter hole halfway through the arm at each mark. Dab some glue into the holes and insert dowels. They should protrude by about 3/8 inch. To use the jig, position it on the workpiece so that the dowels butt against opposite faces of the stock. Holding the jig with one hand, fit the drill bit into the bushing and bore the hole *(above)*.

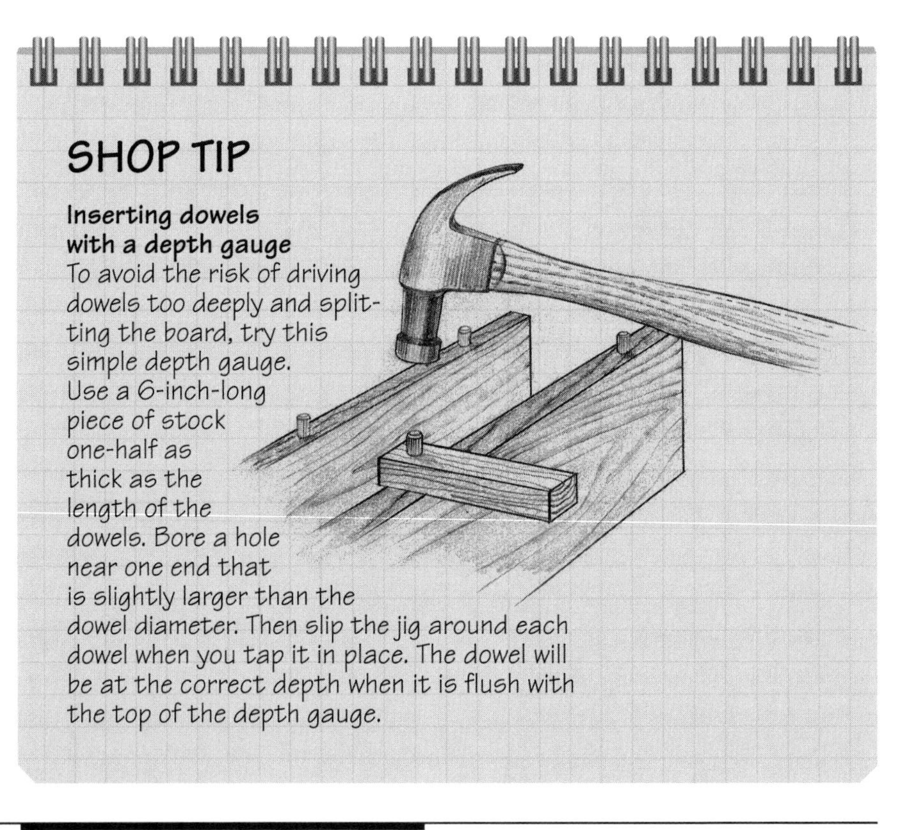

SHOP TIP

Inserting dowels with a depth gauge

To avoid the risk of driving dowels too deeply and splitting the board, try this simple depth gauge. Use a 6-inch-long piece of stock one-half as thick as the length of the dowels. Bore a hole near one end that is slightly larger than the dowel diameter. Then slip the jig around each dowel when you tap it in place. The dowel will be at the correct depth when it is flush with the top of the depth gauge.

TILTING TABLE JIG

1 Making the table
To bore angled holes without tilting the drill press table, use the tilting jig shown at right, built from solid stock and ¾-inch plywood. Refer to the illustration for suggested dimensions. Connect the jig top to the base using a sturdy piano hinge. Cut a ½-inch-wide slot in the support brackets and screw each one to the top; secure the brackets to the base with wing nuts, washers, and hanger bolts.

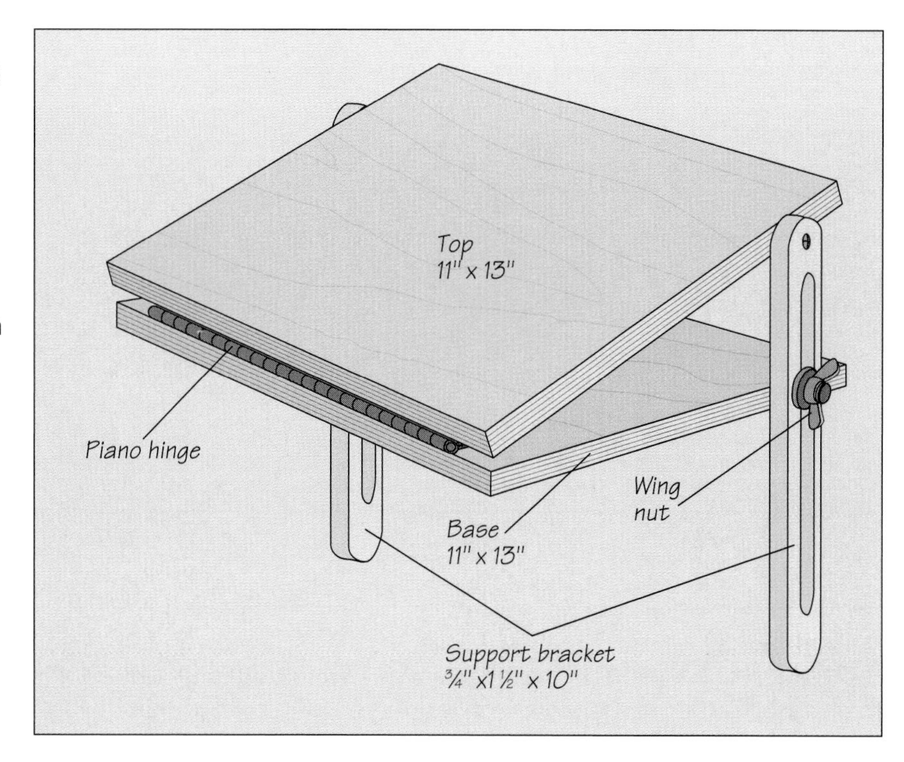

2 Drilling angled holes
Center the jig under the drill press spindle, clamp the base to the table, and loosen the wing nuts. Use a protractor and a sliding bevel to set the angle of the top, then tighten the wing nuts. Clamp the workpiece to the jig and bore the hole *(below)*.

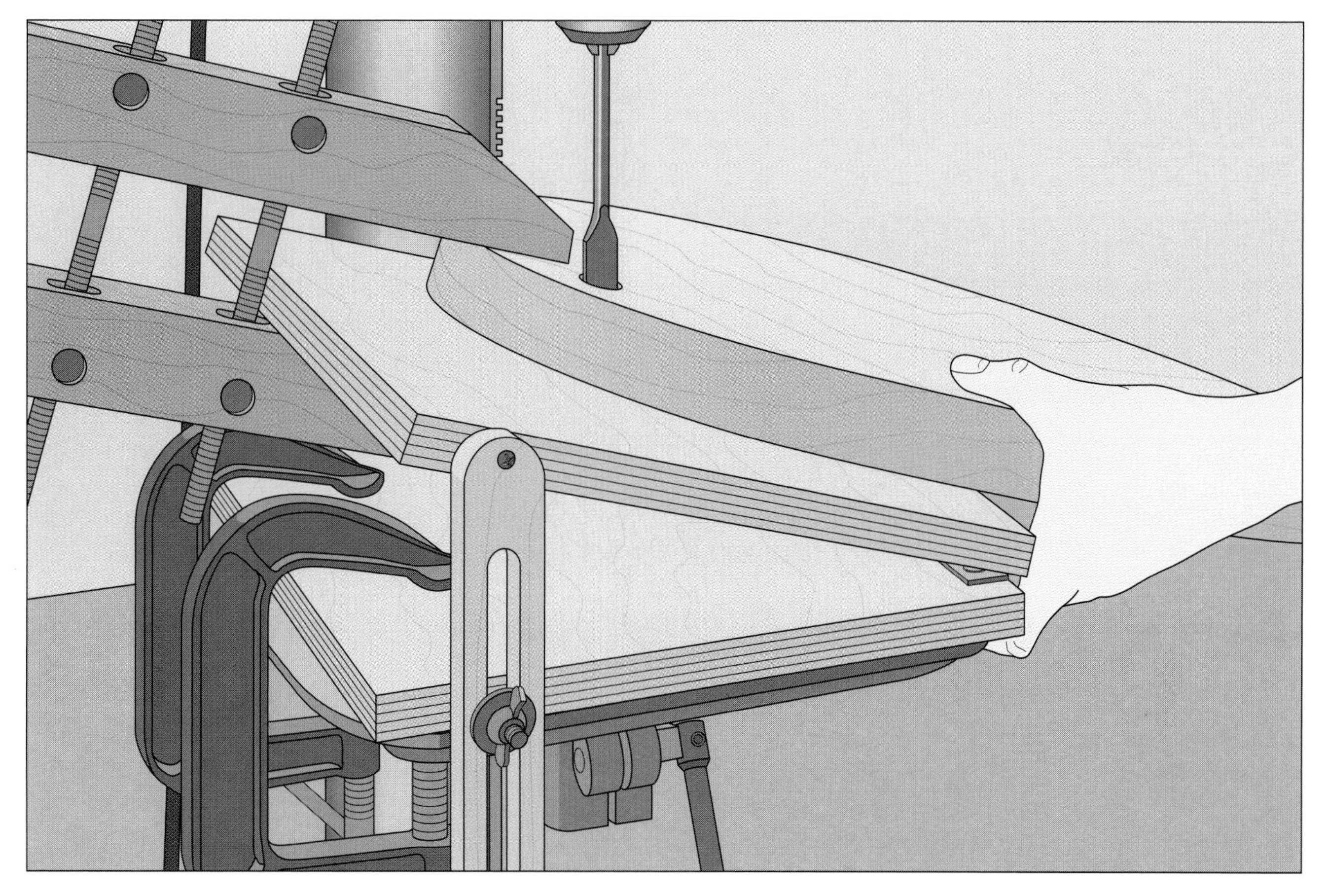

JIG FOR DRILLING EQUALLY SPACED HOLES

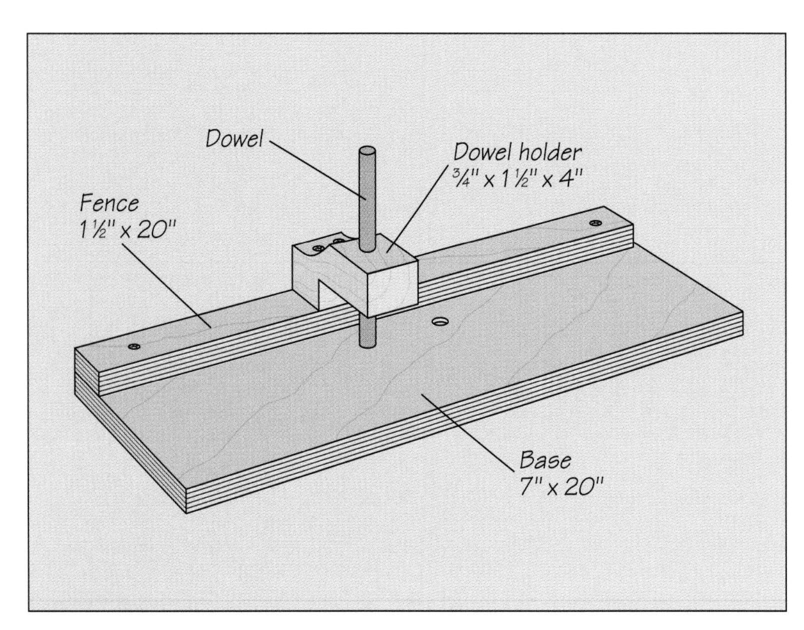

1 Making the jig
Make this jig to simplify the task of boring a row of equally spaced holes on the drill press. Screw the fence flush with one edge of the base, then attach a wood block to the center of the fence as a dowel holder. (The hole for the dowel is made the first time you use the jig.) The dimensions given in the illustration will suit most drill press tables.

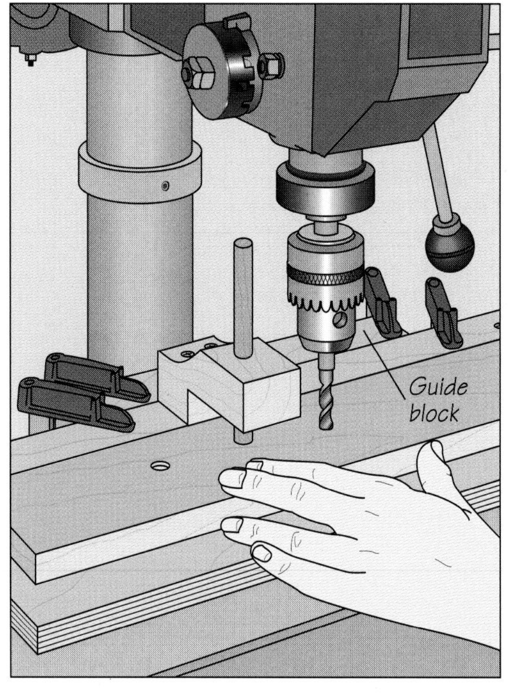

2 Boring the holes
Set the jig on the drill press table. Mark the location of the first two holes and seat the workpiece against the fence of the jig, aligning the first drilling mark under the bit. Butt a guide block against the back of the jig and clamp it to the table. Bore the first hole, slide the jig along the guide block, and bore a hole through the dowel holder. Fit a dowel through the hole in the holder and into the hole in the workpiece. Slide the jig until the second drilling mark is aligned under the bit. Clamp the jig to the table and bore the second hole. To drill each remaining hole, retract the dowel and slide the workpiece along the jig fence. Push the dowel down into the last hole you made.

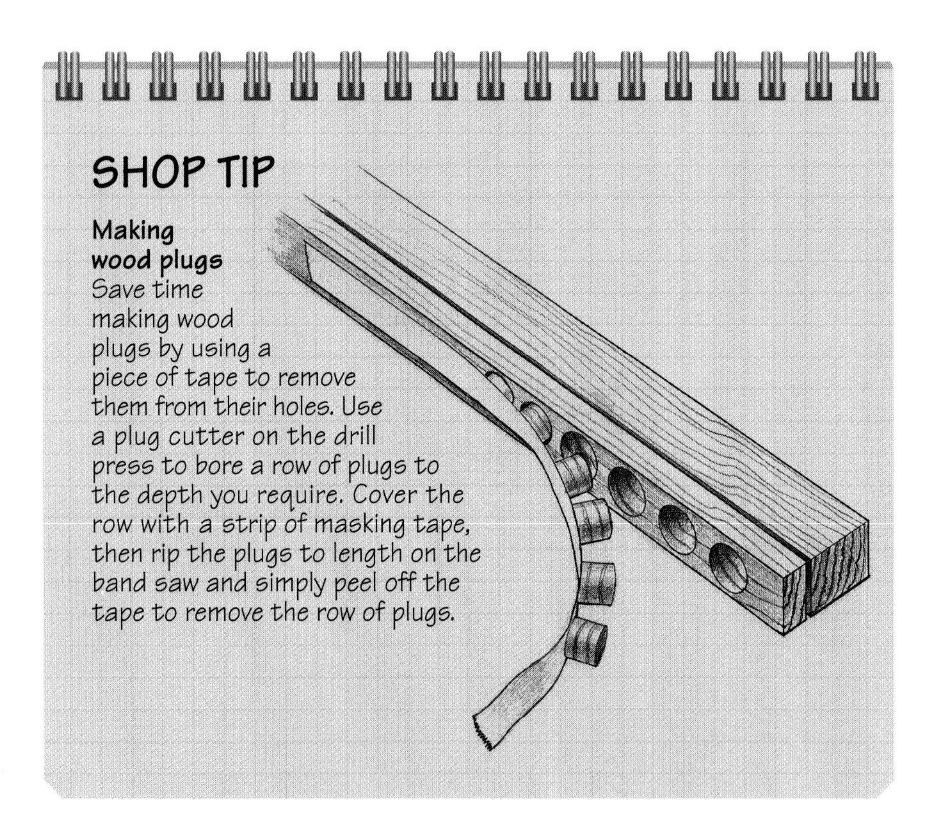

SHOP TIP

Making wood plugs
Save time making wood plugs by using a piece of tape to remove them from their holes. Use a plug cutter on the drill press to bore a row of plugs to the depth you require. Cover the row with a strip of masking tape, then rip the plugs to length on the band saw and simply peel off the tape to remove the row of plugs.

V-BLOCK JIG

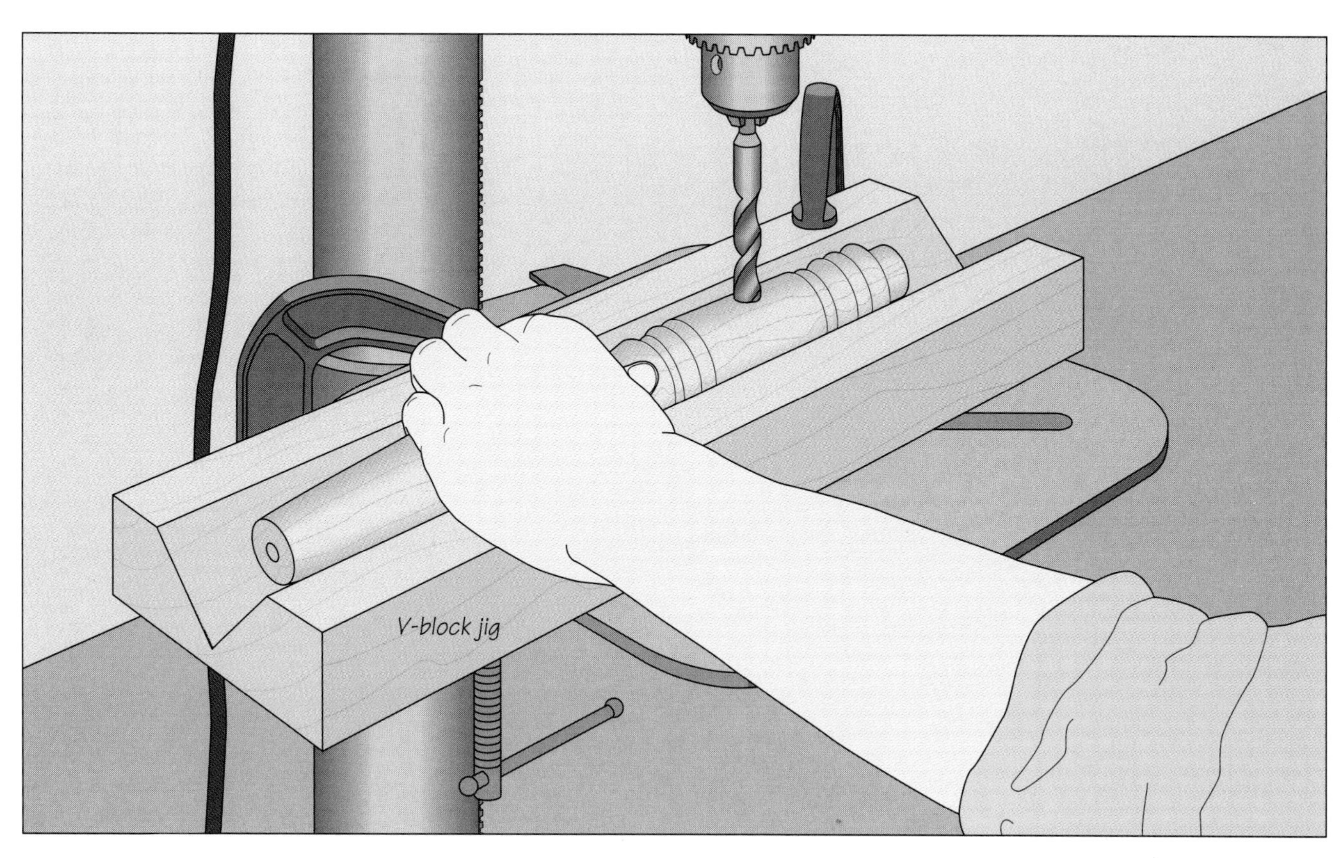

Boring holes in cylindrical stock
To make the simple V-block jig shown above, mark a right-angle V on the end of a piece of solid stock that is long enough to hold your workpiece. Then, adjust the blade angle on your table saw to 45° and align one cutting line with the blade. Butt the rip fence against the stock and feed the board to cut the first side of the V. Reverse the piece and make the second cut. To use the jig, secure it to the drill table so the drill bit touches the center of the V when the quill is extended. Hold the workpiece in the V, aligning the bit over the cutting mark, set the drilling depth, and bore the hole *(above)*.

SHOP TIP

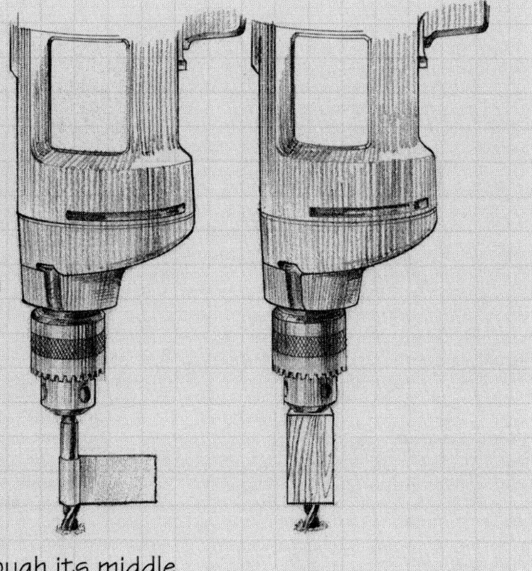

Depth guides for drilling
To bore a hole to an exact depth, use a masking tape flag or a depth stop block. If you are using the tape, measure the drilling depth from the tip of the bit, then wrap a strip of tape around its shank. Withdraw the bit when the tape touches the stock. To use a block, subtract the drilling depth from the length of the bit protruding from the chuck. Cut a piece of 1-by-1 stock to this length, then bore a hole through its middle. Slip the bit through the block and bore your hole until the block touches the workpiece.

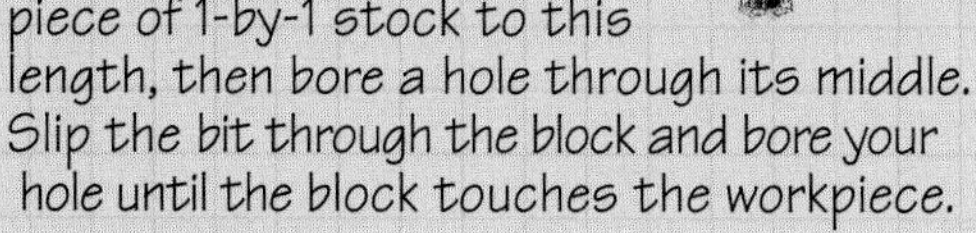

POCKET HOLE JIG

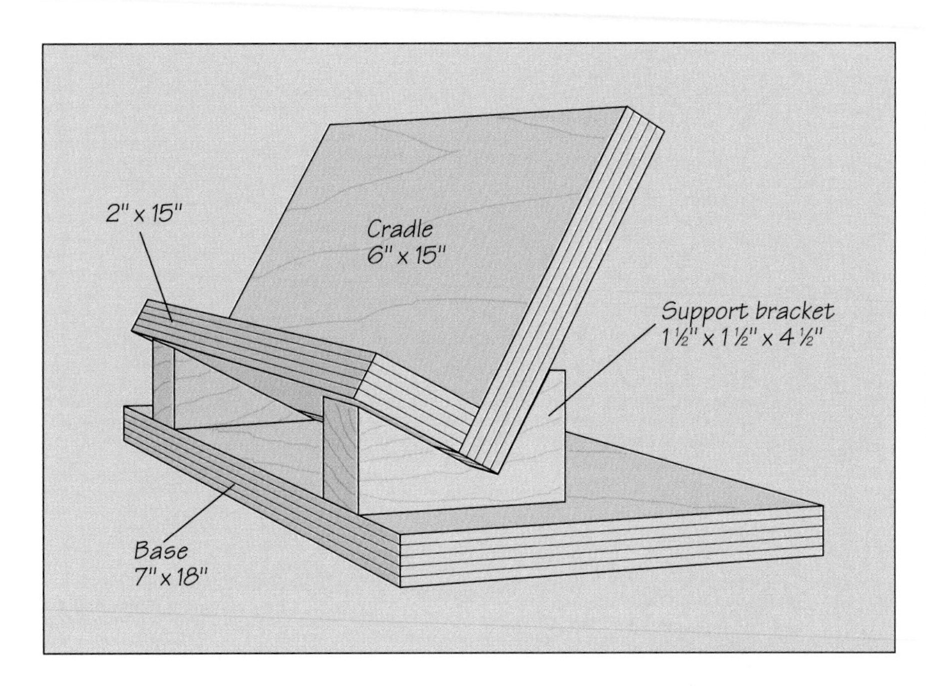

1 Making the jig
To bore pocket holes on the drill press, use this jig made from ¾-inch plywood and two small pieces of solid stock. Refer to the illustration at left for suggested dimensions. Screw the two sides of the cradle together to form an L. Then cut a 90° angle wedge from each support bracket so that the wide side of the cradle will sit at an angle of about 20° from the vertical. Screw the brackets to the jig base and attach the cradle to the brackets.

2 Drilling pocket holes
Seat the workpiece in the cradle with the side to be drilled facing out and the top edge sitting in the V of the cradle. Bore the holes in two steps with two different bits: a Forstner bit slightly larger than the diameter of the screw heads, so they can be recessed, and a brad-point bit a little larger than the screw shanks to allow for wood movement. Install the brad-point bit in the chuck and position the jig on the drill press table so the bit aligns with the center of the bottom edge of the workpiece *(inset)*. Clamp the jig to the table and replace the brad-point bit with the Forstner. Holding the workpiece firmly in the jig, feed the bit slowly to bore each hole just deep enough to recess the screw heads *(right)*. To complete the pocket holes, reinstall the brad-point bit and bore through the workpiece.

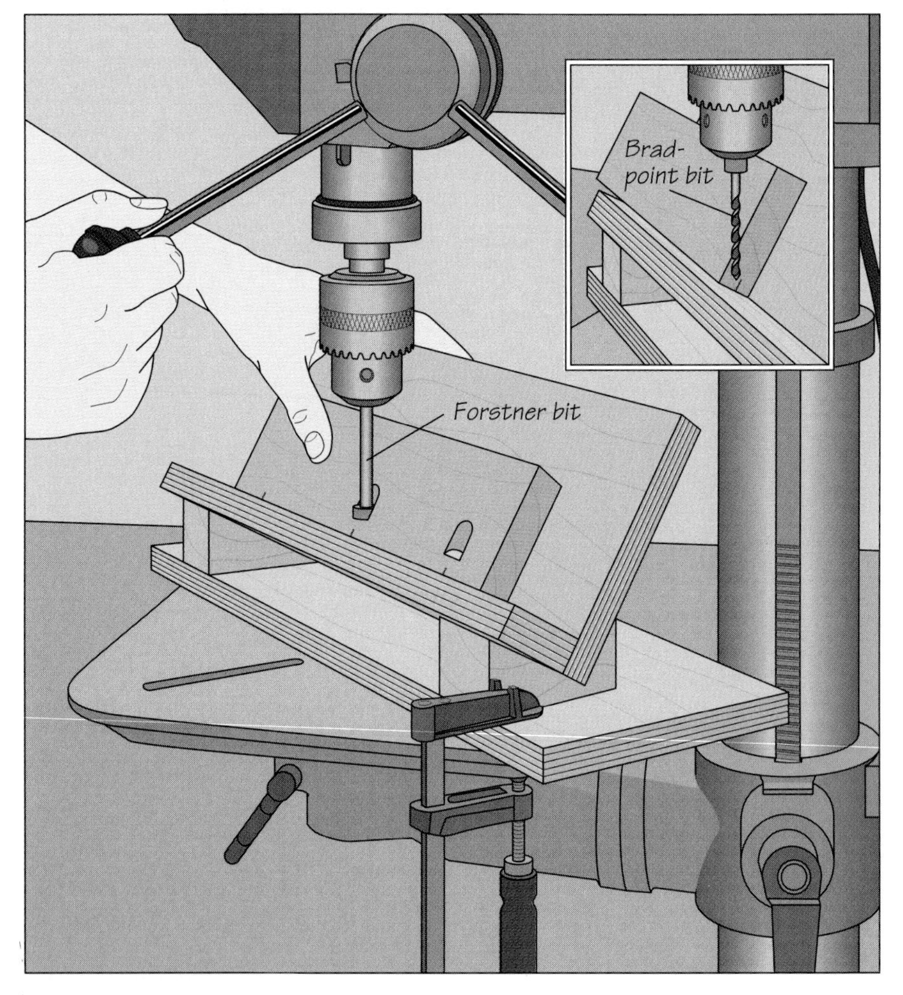

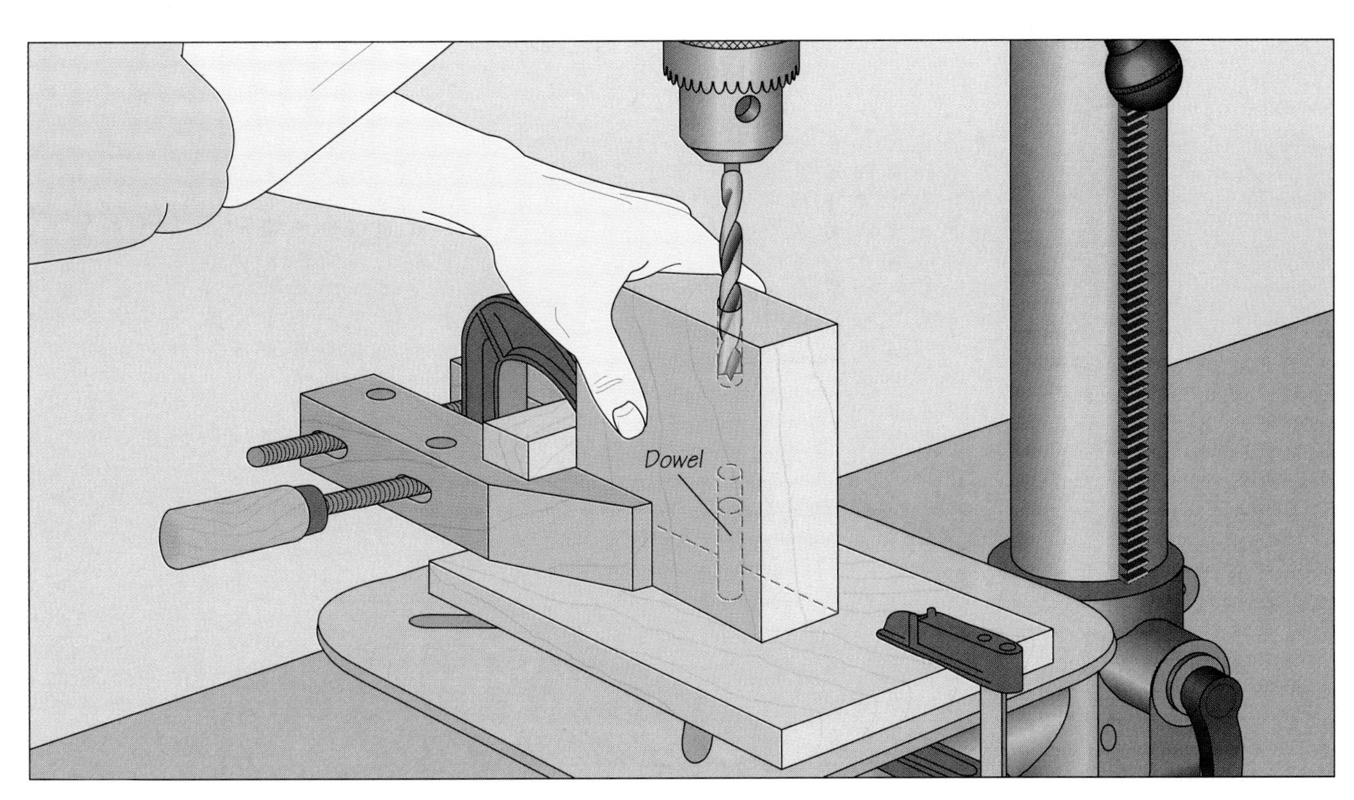

SHOP TIP

Cutting rosettes on the drill press

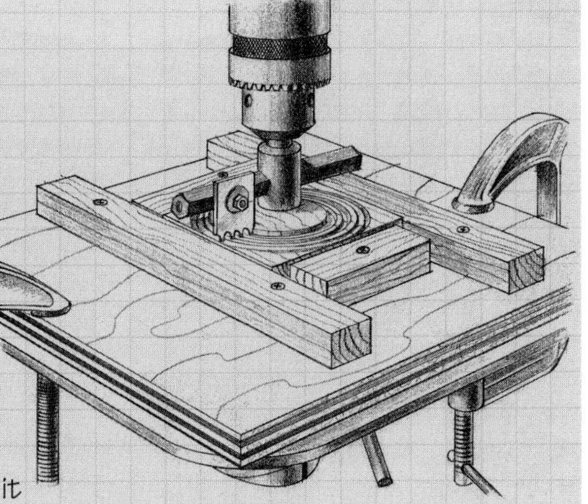

You can cut decorative rosettes by modifying a drill press fly cutter with a beading blade from a table saw molding head. Notch the fly cutter arm to accommodate the beading blade, locating the cutter about 1 inch from the end of the arm. Make sure it fits securely in the notch so it cannot shift during use. Bore a hole through the arm and use a bolt, washer, and nut to fasten the blade in the notch with its flat face toward the direction of cutter rotation. Set your workpiece on a wood base with stop blocks to hold it securely. Clamp the setup to the drill press table, aligning the center of the fly cutter over the center of the stock. Turn on the drill and lower the quill until the blade lightly contacts the wood. Continue cutting until the rosette has the profile you want.

Exceeding a drill press's quill stroke

The maximum distance that the quill of a drill press can be extended is called its stroke. On most tools, this is no more than 4 inches. To drill a deeper hole, you may use an extension bit or, if the hole and the workpiece measure less than twice the quill stroke, perform the operation in two stages with a standard bit. A simple jig solves any alignment problems. First, clamp a scrap board to the drill press table and bore a guide hole. Then, clamp the workpiece to the board and bore into it as deeply as the quill stroke will allow. Remove the workpiece, fit a dowel into the guide hole, slip the workpiece over the dowel, and finish boring the hole from the other side *(above)*. The dowel will ensure that the two holes are perfectly aligned.

TURNING JIGS

Lathes enable a woodworker to turn blocks of wood of almost any irregular shape into beautiful, rounded creations. Woodturners often speak of how visual their art is; they see a chair leg, bowl, or pepper mill seem to grow from a spinning blank. Yet despite its visual element and the emphasis on "feel" for judging the progress of a turned workpiece, turning also depends on accuracy and careful measurement.

The jigs in this chapter will help you turn a workpiece with precision. Use one of the setup jigs shown on pages 70 and 71 to mount a blank properly on the lathe, even if the workpiece is irregularly shaped. A layout jig *(page 73)* is ideal for quickly and accurately outlining the location of the contours on a blank, while measuring jigs like the diameter gauge *(page 73)* or bowl depth gauge *(page 77)* come in handy for checking a workpiece's dimensions.

The lathe can also be used for purposes other than turning; see the sanding drum *(page 77)* and the jig for routing flutes in columns *(page 75)*.

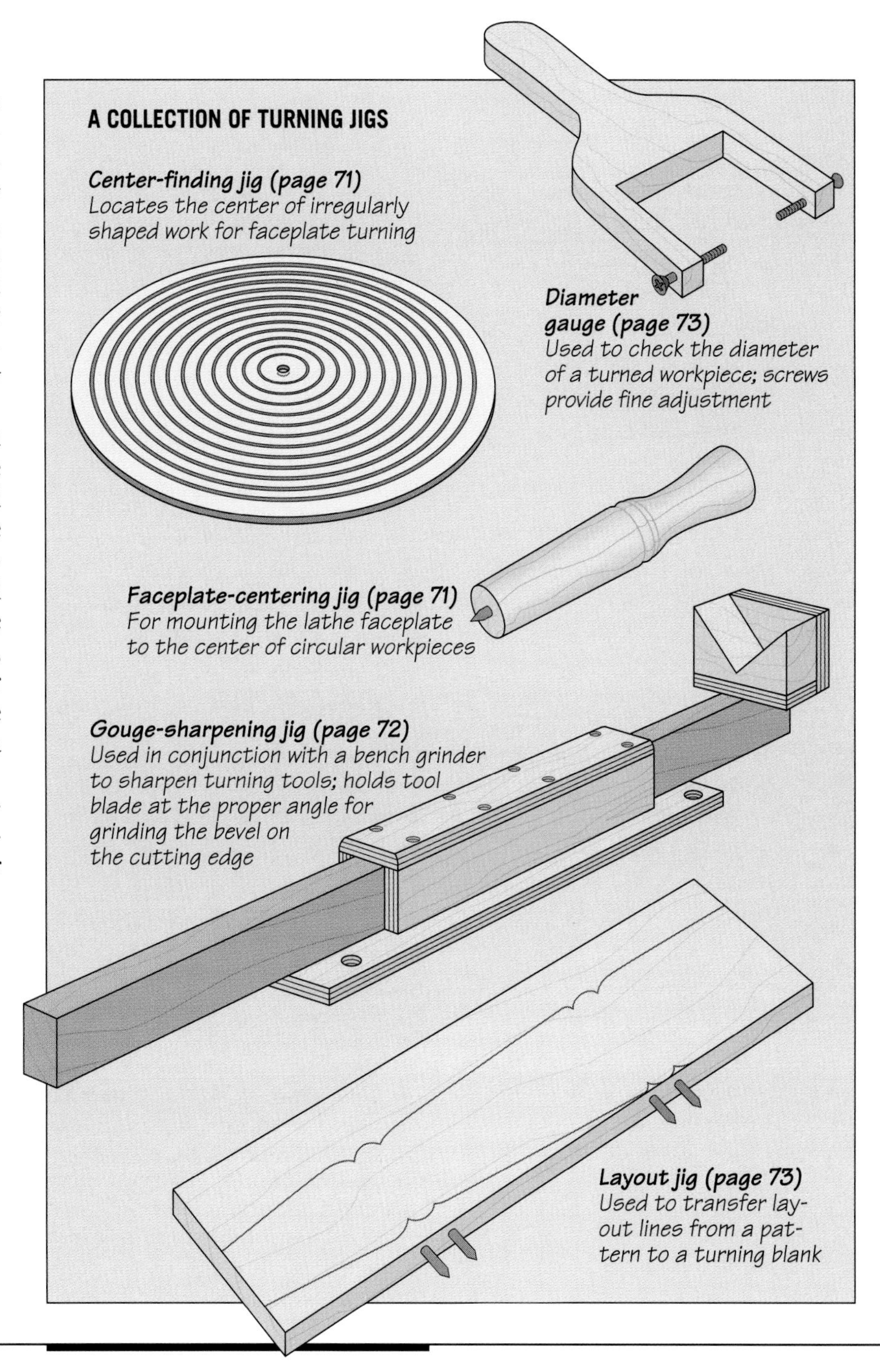

Mounting hollowed-out workpieces on a lathe can be challenging. For the cocobolo bud vase being sanded at left, a conical bull-nose tailstock was positioned between the lathe's metal tailstock and the workpiece, allowing the vase to be supported snugly without damage.

A COLLECTION OF TURNING JIGS (continued)

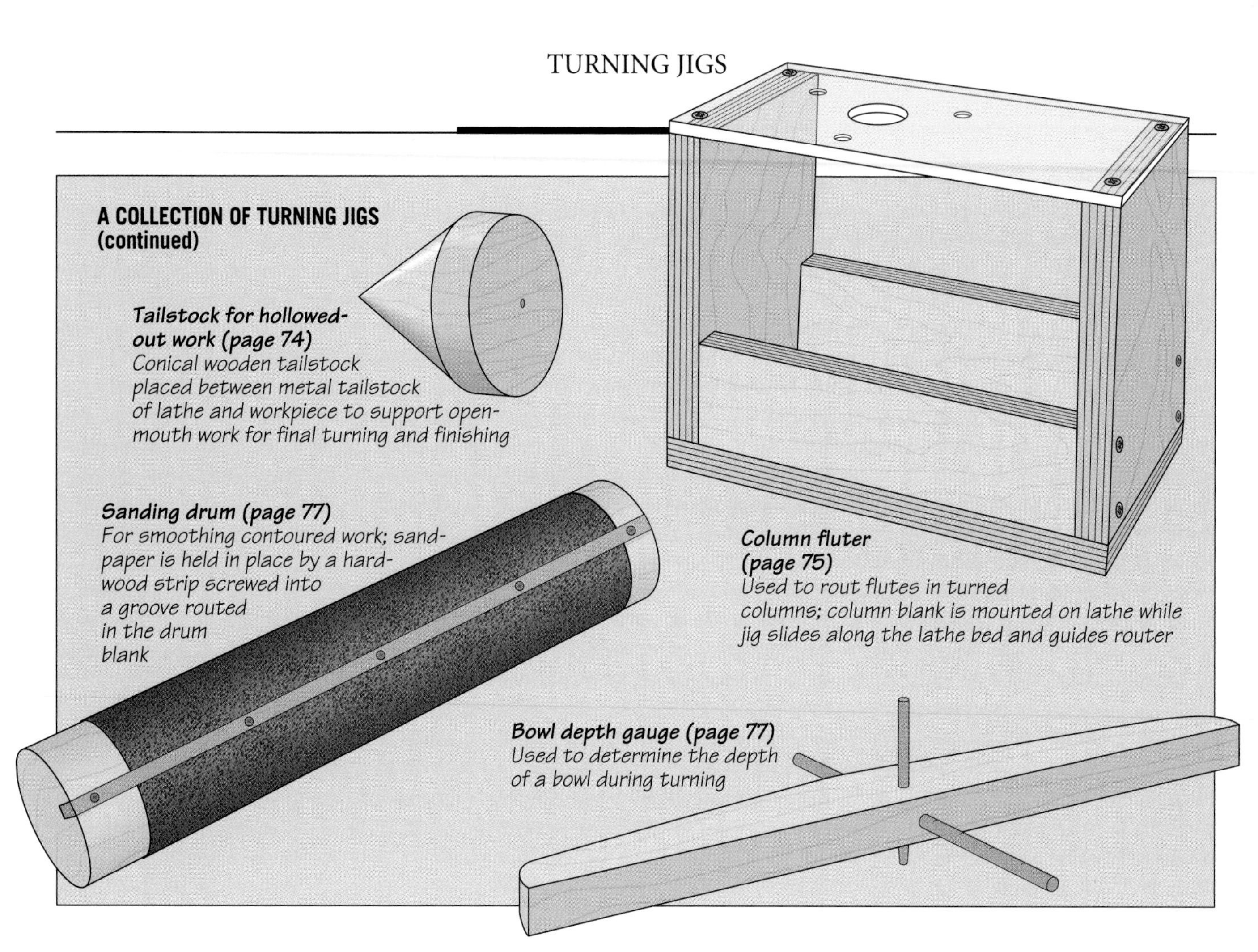

Tailstock for hollowed-out work (page 74)
Conical wooden tailstock placed between metal tailstock of lathe and workpiece to support open-mouth work for final turning and finishing

Sanding drum (page 77)
For smoothing contoured work; sandpaper is held in place by a hardwood strip screwed into a groove routed in the drum blank

Column fluter (page 75)
Used to rout flutes in turned columns; column blank is mounted on lathe while jig slides along the lathe bed and guides router

Bowl depth gauge (page 77)
Used to determine the depth of a bowl during turning

SHOP TIP

Centering a spindle blank

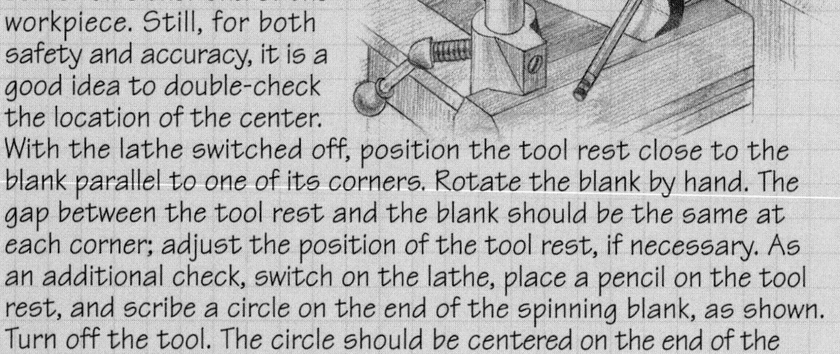

Finding the center of a blank for spindle turning is traditionally done by marking two diagonal lines from corner to corner on either end of the workpiece. Still, for both safety and accuracy, it is a good idea to double-check the location of the center. With the lathe switched off, position the tool rest close to the blank parallel to one of its corners. Rotate the blank by hand. The gap between the tool rest and the blank should be the same at each corner; adjust the position of the tool rest, if necessary. As an additional check, switch on the lathe, place a pencil on the tool rest, and scribe a circle on the end of the spinning blank, as shown. Turn off the tool. The circle should be centered on the end of the blank; if not, adjust the tailstock and repeat.

A FACEPLATE-CENTERING JIG

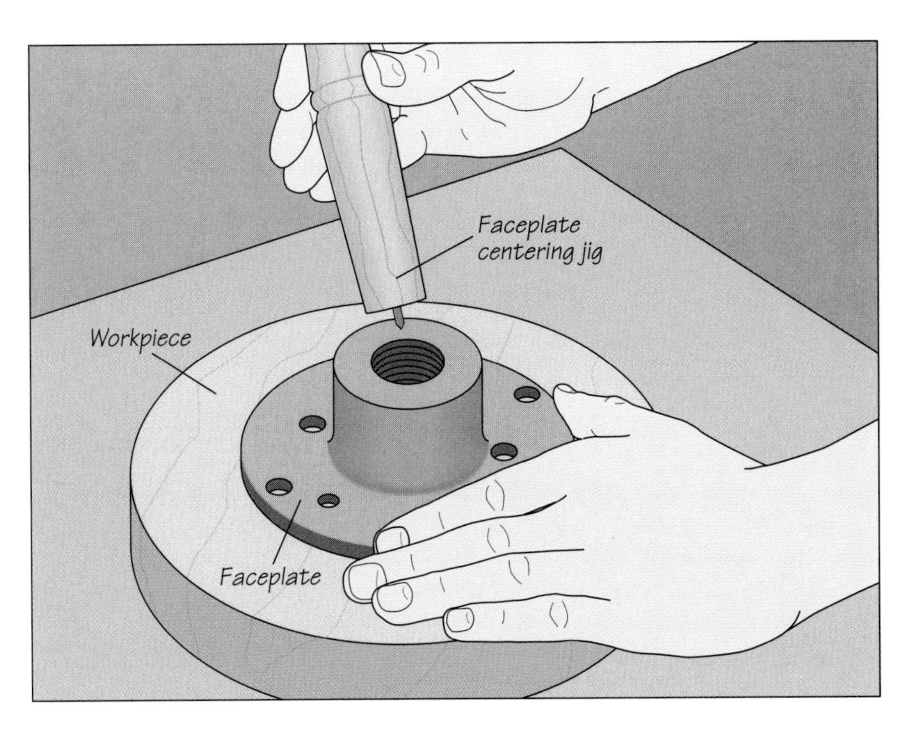

Centering the faceplate on a workpiece
To center your lathe's faceplate on a circular workpiece, use the handy jig shown at right. Turn a cylindrical piece of wood to the diameter of the faceplate's threaded hole, tapering the end slightly. (You may wish to form a handle at the top end of the jig.) Drive a nail into the center of the tapered end, cut off the nailhead, and grind a sharp point. To use the jig, mark the center of the workpiece with an awl *(page 132)*. Next, set the faceplate on the workpiece with the awl mark in the middle of the threaded hole, insert the jig in the hole as shown, and "feel" for the mark with the nail tip. Holding the jig in place, screw the faceplate to the workpiece.

A CENTER-FINDING JIG

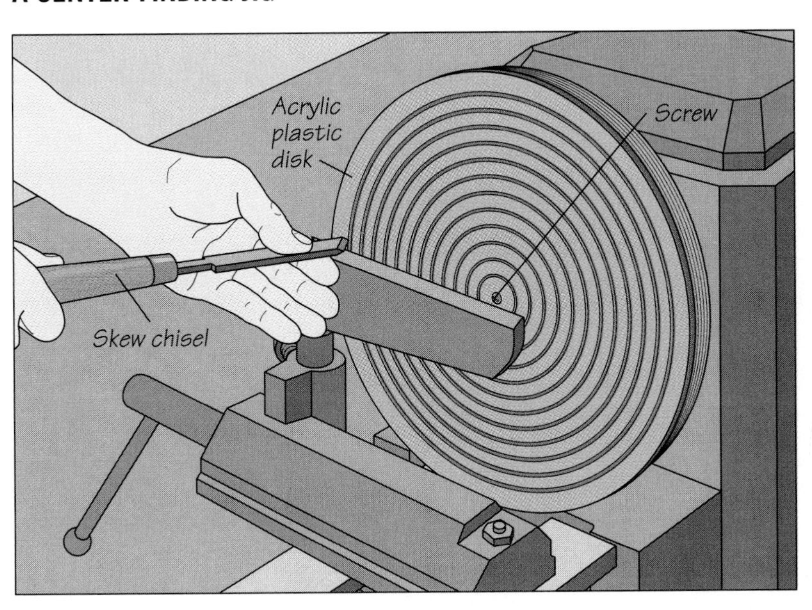

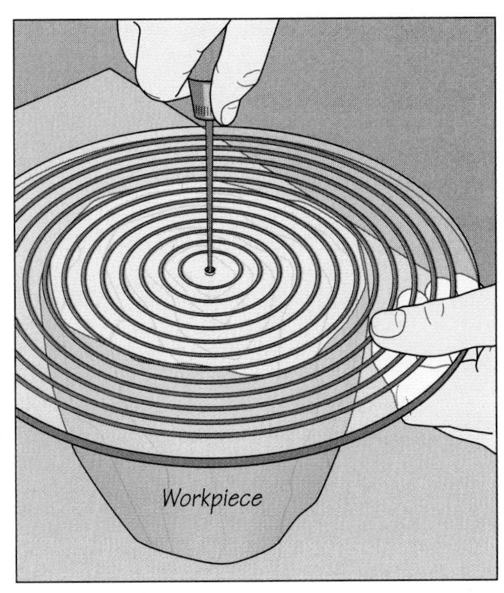

Attaching the faceplate to an irregularly shaped workpiece
The jig shown above will enable you to center the faceplate on an irregularly shaped workpiece. Cut pieces of 1/4-inch clear acrylic plastic and 3/4-inch plywood into 12-inch-diameter disks. Attach the two together with double-sided tape and a screw. Mount the assembly to your faceplate so the plastic is facing out, then use a skew chisel to cut a series of equally spaced, concentric rings into the disk *(above, left)*. Remove the plastic disk and spray it with black paint. Once the paint has dried, peel the backing paper off both sides of the plastic. To use the jig, set it on the workpiece so as much of it as possible is within one of the rings and mark the center with an awl *(above, right)*. Use the mark as a centerpoint for mounting the faceplate.

GOUGE-SHARPENING JIG

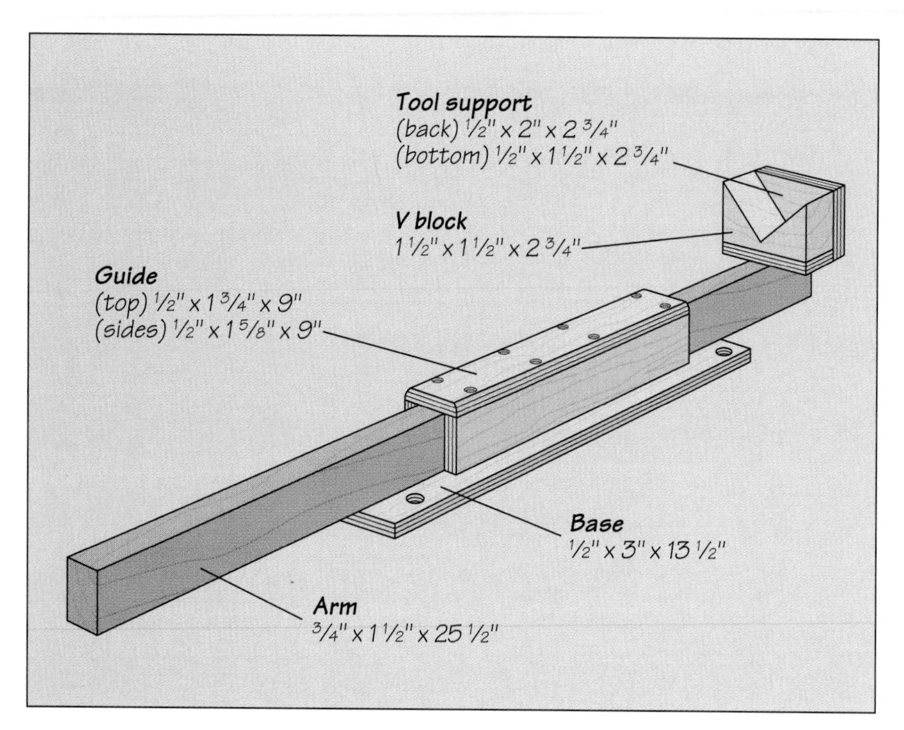

1 Making the jig
The jig shown at left guarantees that the tip of a turning gouge will contact the wheel of your bench grinder at the correct angle to restore the bevel on the cutting edge. The dimensions in the illustration will accommodate most gouges. Cut the base and the guide from ½-inch plywood; screw the guide together and fasten it to the base with screws countersunk from underneath. Make sure the opening created by the guide is large enough to allow the arm to slide through freely. Cut the arm from 1-by-2 stock and the tool support from ½-inch plywood. Screw the two parts of the tool support together, then fasten the bottom to the arm, flush with one end. For the V block, cut a small wood block to size and saw a 90° wedge out of one side. Glue the block to the tool support.

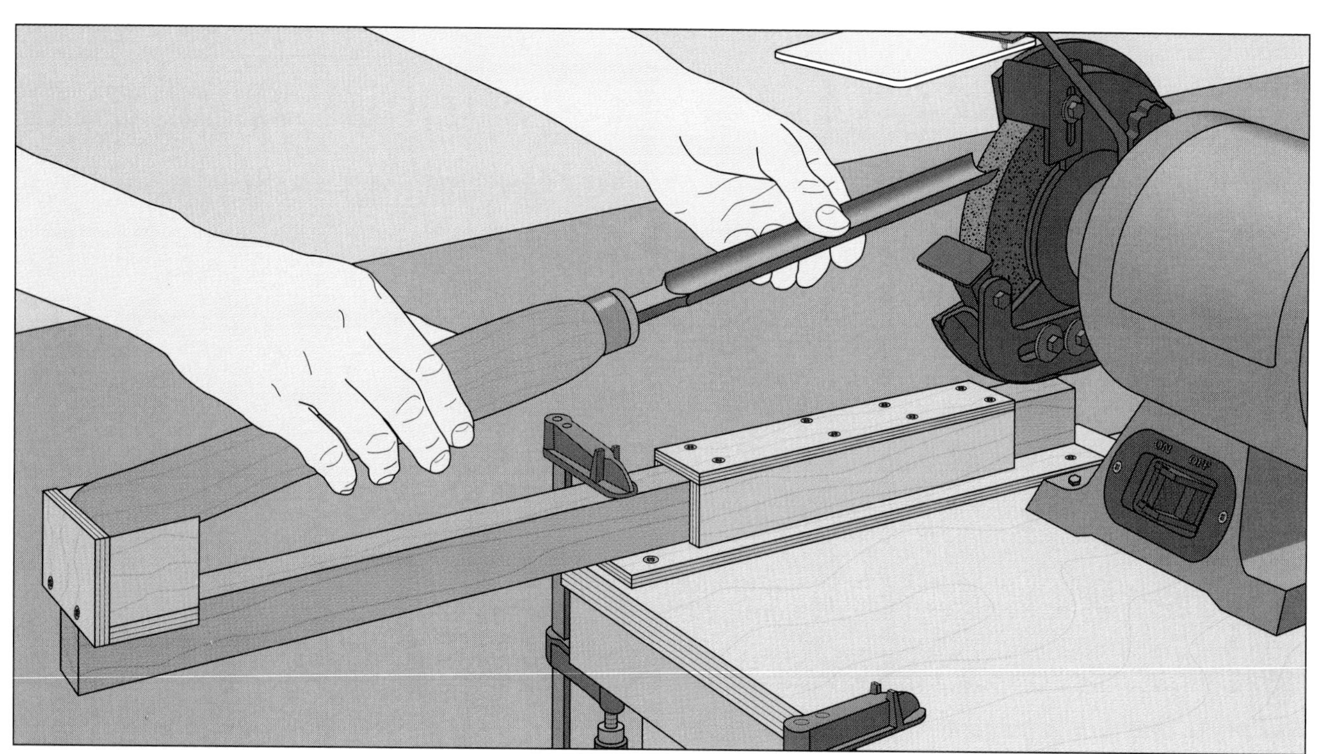

2 Sharpening a gouge
Set the jig on a work surface so the arm lines up directly under the grinding wheel. Seat the gouge handle in the V block and slide the arm so the beveled edge of the gouge rests flat on the grinding wheel. Clamp the arm in place. Then, with the gouge clear of the wheel, switch on the grinder and reposition the tool in the jig. Holding the gouge with both hands, rotate it from side to side so the beveled edge runs across the wheel *(above)*. Check the cutting edge periodically until the bevel is fully formed.

A DIAMETER GAUGE AND A LAYOUT JIG

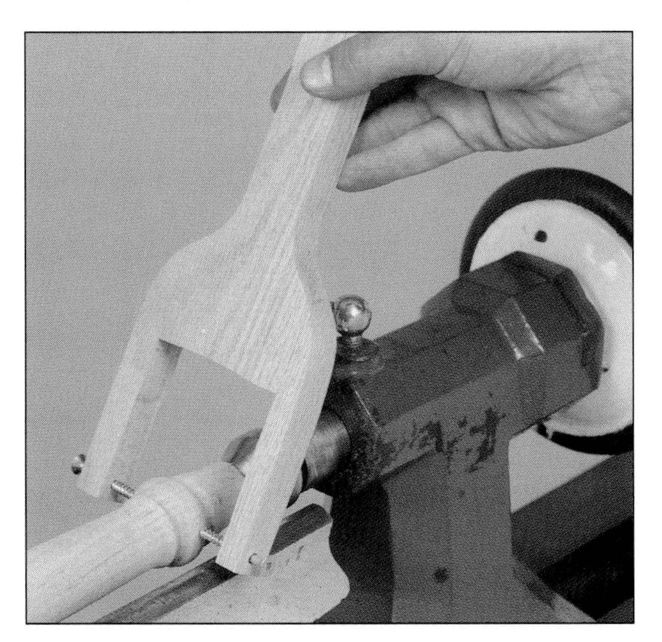

Resembling a tuning fork, the diameter gauge shown above can be used like calipers to check the diameter of turned workpieces. The jig is cut from solid wood; the distance between the arms should be slightly wider than the largest finished diameter to be turned. Two screws driven into the inside of the arms can be adjusted to set precise measurements.

Making and using a layout jig for spindle turning
The layout jig shown above makes it easier to turn multiple copies of spindle work such as table legs by allowing you to scribe layout lines in exactly the same location on every blank. Trace your design on a piece of scrap about the same length as your blanks and wider than their diameter. Drive a nail into the edge of the jig at each point where a transitional element of the design—such as a bead or fillet—begins and ends. Snip off the nail heads and grind the nails to sharp points. Once you have roughed out a cylinder, simply press the jig into the spinning blank; the nails will score the layout lines *(above)*.

SHOP TIP

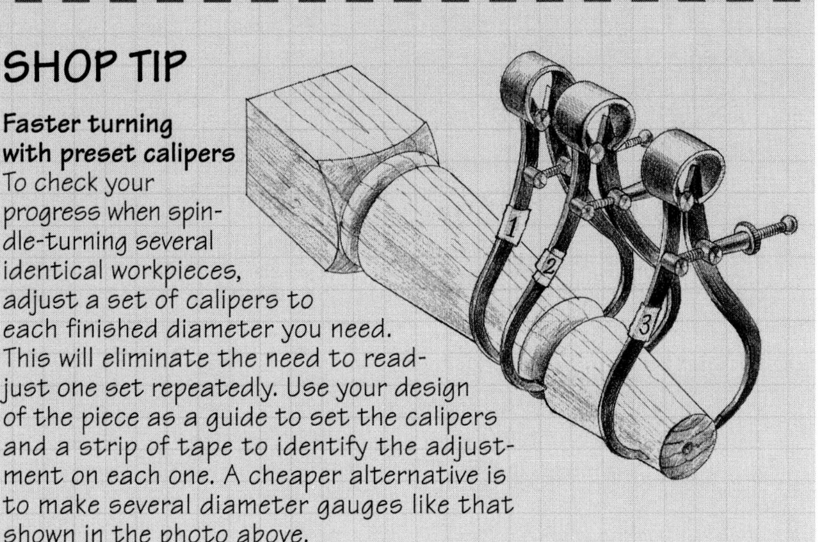

Faster turning with preset calipers
To check your progress when spindle-turning several identical workpieces, adjust a set of calipers to each finished diameter you need. This will eliminate the need to readjust one set repeatedly. Use your design of the piece as a guide to set the calipers and a strip of tape to identify the adjustment on each one. A cheaper alternative is to make several diameter gauges like that shown in the photo above.

A TAILSTOCK FOR HOLLOWED-OUT WORK

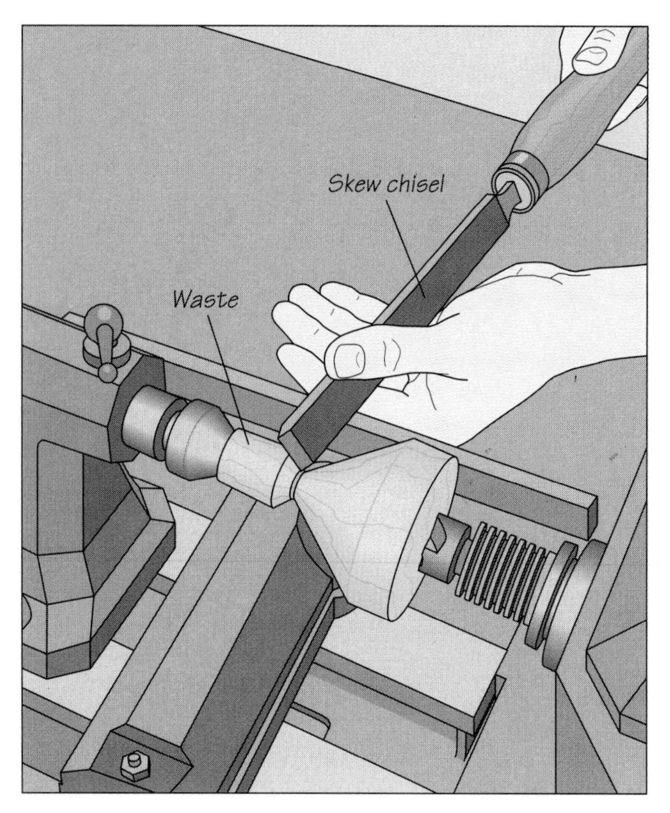

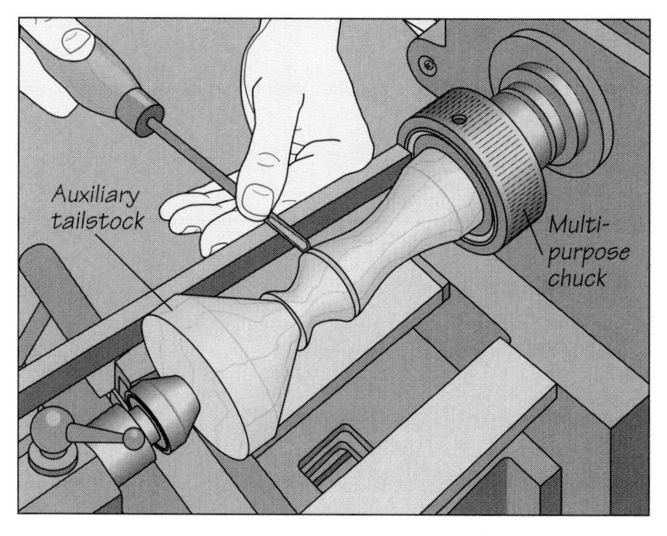

2 Turning hollowed-out work
Mount the workpiece to a multi-purpose chuck and secure the chuck to the lathe. Turn the inside of the work. Once you are ready to turn the outside, insert the tapered end of the jig into the opening in the work and, holding the auxiliary tailstock in place, advance the machine's tailstock until it contacts the jig and holds the jig and work securely. Tighten the tailstock and finish the workpiece *(above)*.

1 Making the tailstock
To support hollowed-out workpieces on the lathe, use an auxiliary tailstock like the one shown above. Turned from hardwood, this simple jig supports the workpiece at the rim only, and rotates along with the lathe's tailstock, preventing vibration and burning of the workpiece. To make the device, first turn a cylinder from a blank about 4 inches in length, then shape the cone with a roughing gouge and use a skew chisel to smooth the cone and separate it from the waste *(above)*. The size and taper of the tailstock will vary according to the size and diameter of the workpiece; a cone 3½ inches wide at the base with a 60° taper suits most small work.

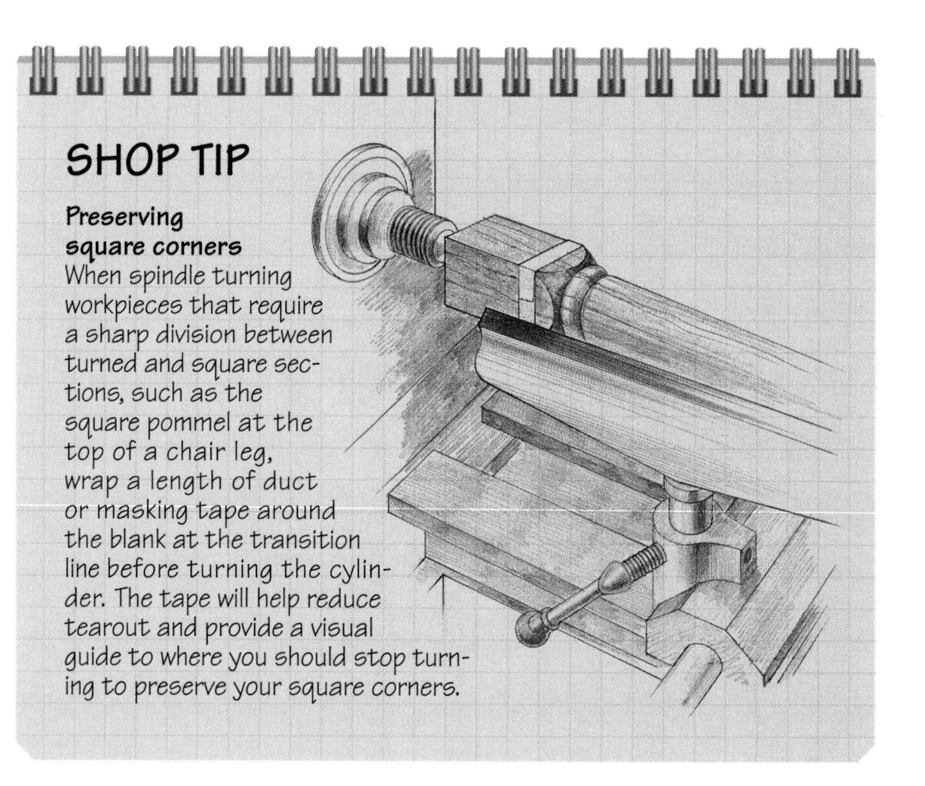

SHOP TIP

Preserving square corners
When spindle turning workpieces that require a sharp division between turned and square sections, such as the square pommel at the top of a chair leg, wrap a length of duct or masking tape around the blank at the transition line before turning the cylinder. The tape will help reduce tearout and provide a visual guide to where you should stop turning to preserve your square corners.

JIG FOR FLUTING COLUMNS

1 Building the jig
With the box-like jig shown at right you can rout flutes in a column while it is mounted in the lathe. Cut the parts of the jig from ¾-inch plywood, except for the top, which is made from ¼-inch clear acrylic. Refer to the illustration for suggested dimensions; the jig should be long and wide enough to support the router and high enough so the top just clears the column when the jig rests on the lathe bed. Once the top, bottom, and sides are assembled, add two braces to make the jig more rigid. Install a double-bearing piloted fluting bit in your router, drill a bit clearance hole through the jig top, and screw the tool's base plate to the jig. The router should be positioned so the bit will lie alongside the column when the jig is used. Be sure all tools are unplugged during setup.

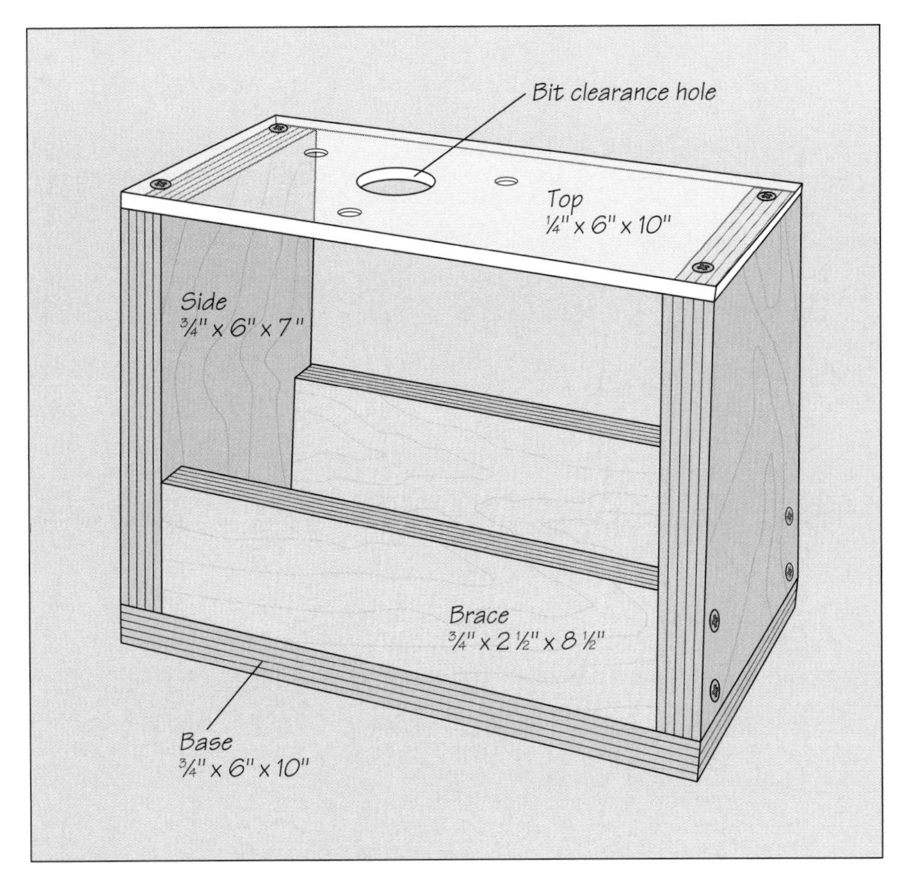

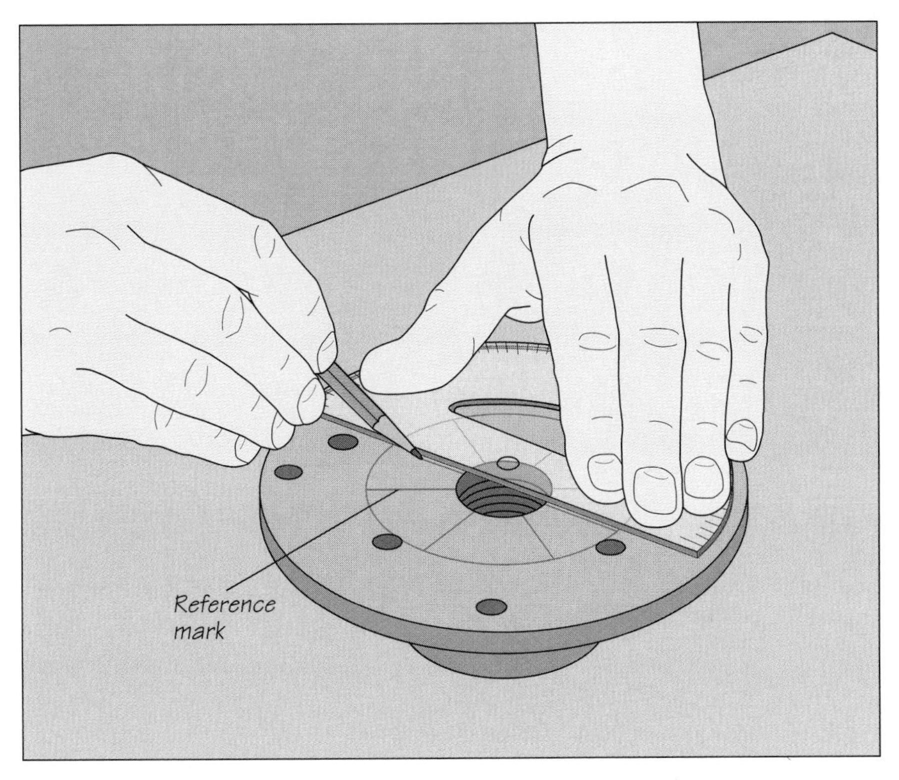

2 Preparing the column and the lathe
To ensure that the flutes are spaced equally around the column, mark cutting lines for the flutes on the blank—in this example, 12 in all. Make corresponding reference marks on the lathe faceplate by dividing its circumference, or 360°, into 12 equal segments spaced 30° apart. Mark the lines with a pencil and a protractor *(left)*. Set the jig on the lathe bed, mount the faceplate on the lathe so that one of the reference marks is in the 12 o'clock position, and mount the column so a cutting mark is at the 3 o'clock position. Tighten a handscrew around the lathe drive shaft to prevent it from rotating and clamp stop blocks to the lathe bed to make sure that all flutes will be the same length.

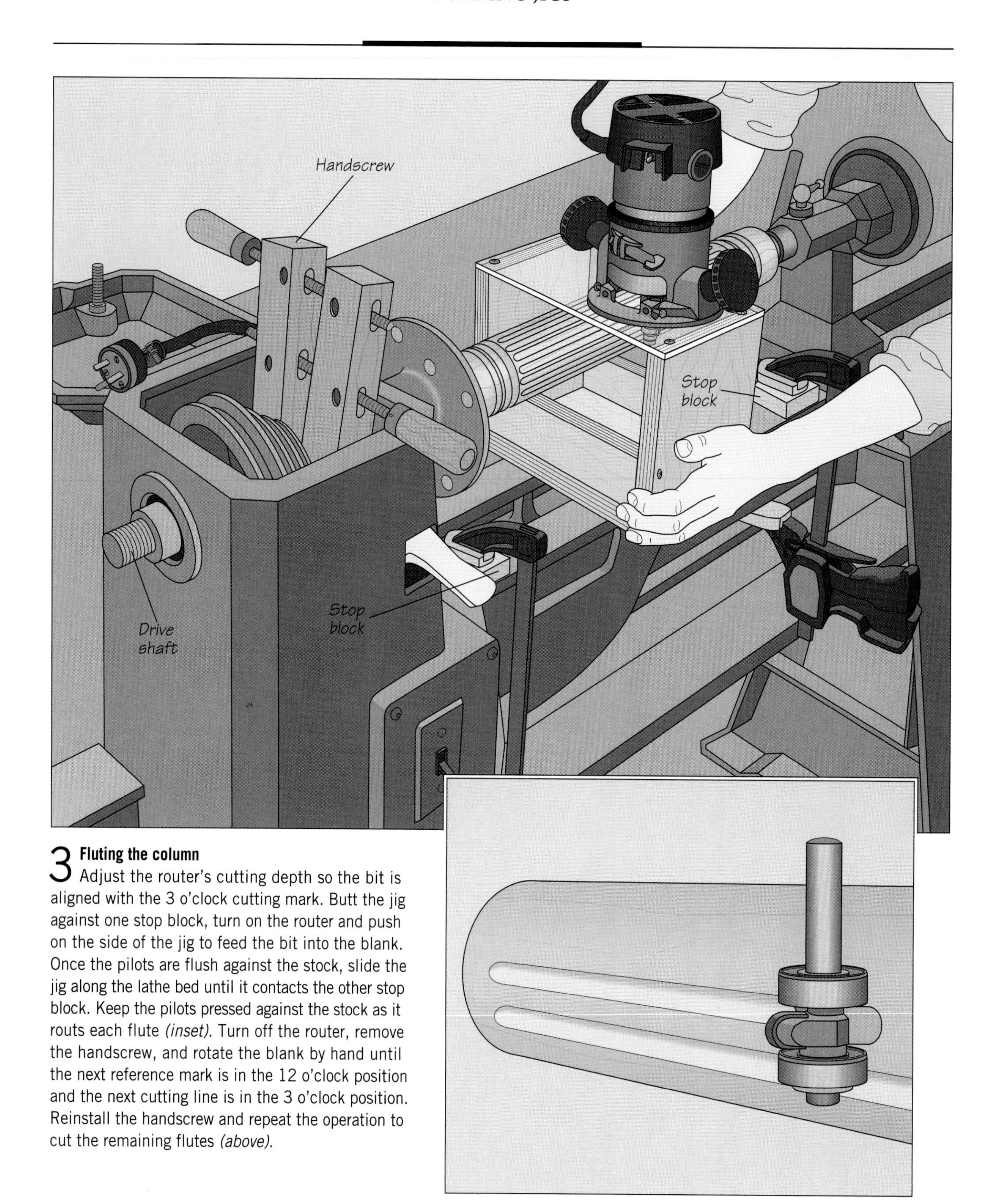

3 Fluting the column
Adjust the router's cutting depth so the bit is aligned with the 3 o'clock cutting mark. Butt the jig against one stop block, turn on the router and push on the side of the jig to feed the bit into the blank. Once the pilots are flush against the stock, slide the jig along the lathe bed until it contacts the other stop block. Keep the pilots pressed against the stock as it routs each flute *(inset)*. Turn off the router, remove the handscrew, and rotate the blank by hand until the next reference mark is in the 12 o'clock position and the next cutting line is in the 3 o'clock position. Reinstall the handscrew and repeat the operation to cut the remaining flutes *(above)*.

JIGS FOR SANDING AND CHECKING DEPTH

Accurately determining depth is essential to faceplate turning; too deep a cut can ruin a bowl. Made from solid wood stock, the bowl-depth gauge shown at right features a pair of perpendicular ¼-inch dowels; the holes for the dowels overlap so that the longer dowel can be adjusted and wedged in place by the shorter, tapered one. Marking depth increments on the long dowel will speed setup.

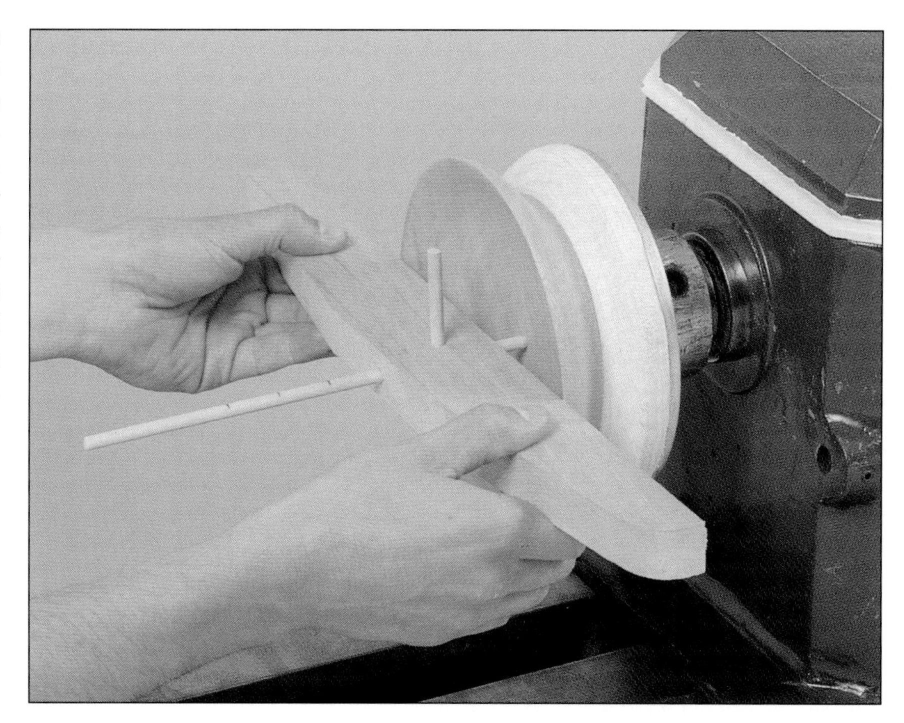

A SANDING DRUM FOR THE LATHE

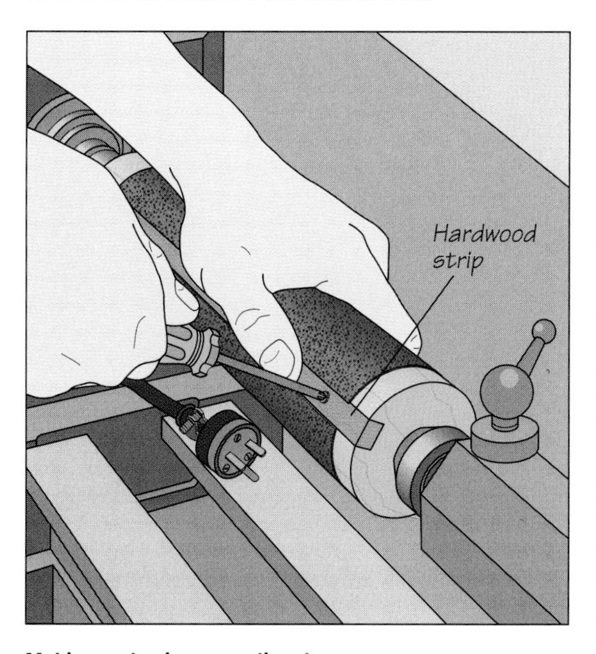

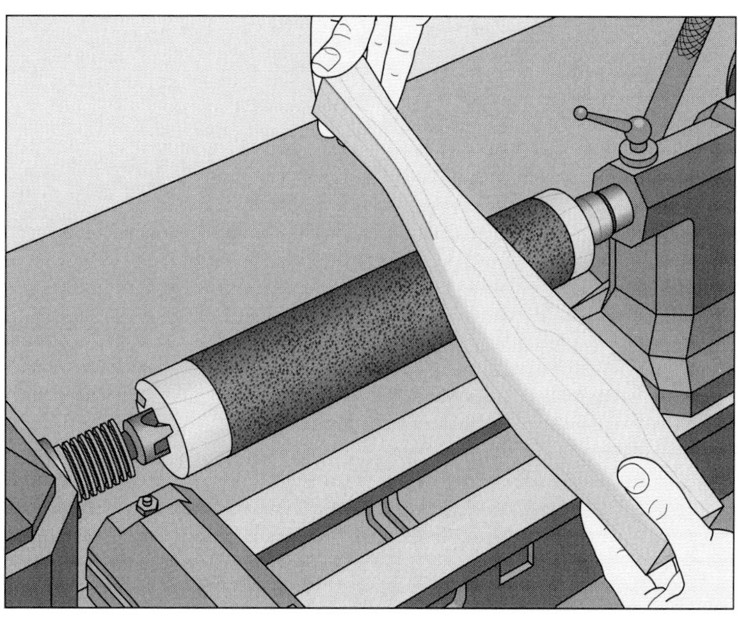

Making and using a sanding drum

The jig shown above will turn your lathe into a powerful drum sander. Start by routing a ½-inch-wide, ⅜-inch-deep groove along the center of one side of a square blank about 14 inches in length. Mount the blank on the lathe and turn it into a cylinder 3½ inches in diameter. Next, cut a sheet of sandpaper to wrap around the cylinder; its ends should overlap at the bottom of the groove. To hold the sandpaper in place, make a hardwood strip to fit in the groove and screw it in place *(above, left)*. To use the sanding drum, mount it between the headstock and tailstock of the lathe. You can then use the drum to sand workpieces with curved contours, such as a cabriole leg *(above, right)*.

TRADEMARK
Jorgensen
REG.U.S. PAT OFF
MADE IN U.S.A

GLUING AND CLAMPING JIGS

Clamps and bench vises are indispensable to the woodworker. Their very simplicity makes them versatile, but the basic clamp or vise can be made to work better or more easily with the help of a jig. The items shown in this chapter will enable you to get the most from the clamps and vises you already have. Other jigs provide alternatives to commercial devices that may not be the best tool for a specific task.

Gluing boards edge-to-edge to assemble panels is a common step in furnituremaking. The jigs shown starting on page 81 provide ways of keeping your clamping setups flat, square, and stable as the adhesive dries. The wall-mounted glue rack *(page 85)* can save considerable shop space, while the wedged clamping bar *(page 83)* and the jig for edge gluing thin stock *(page 84)* will take the place of bar clamps.

For securing stock to your workbench for sanding or planing, make a temporary bench stop *(page 90)*. To fashion a handy wood-carver's vise, try the jig on page 93 that combines a standard pipe clamp with two wood blocks. All the devices in this chapter are simple to build with only a few materials. The dividends they will pay are well worth the effort and expense.

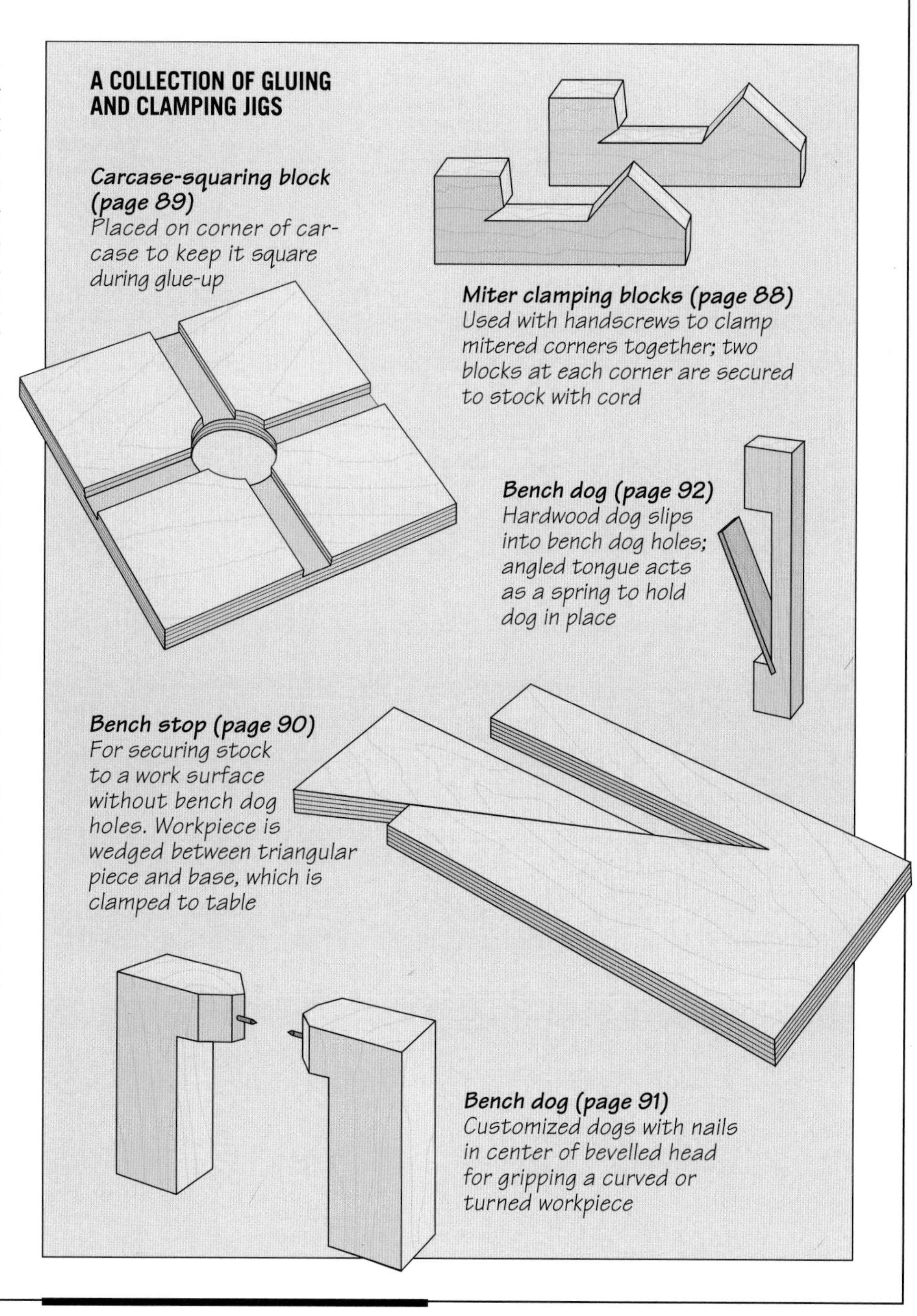

Made from just a few pieces of wood and some hardware, the framing clamp at left allows you to keep the corners of a picture frame square and tightly closed during glue-up.

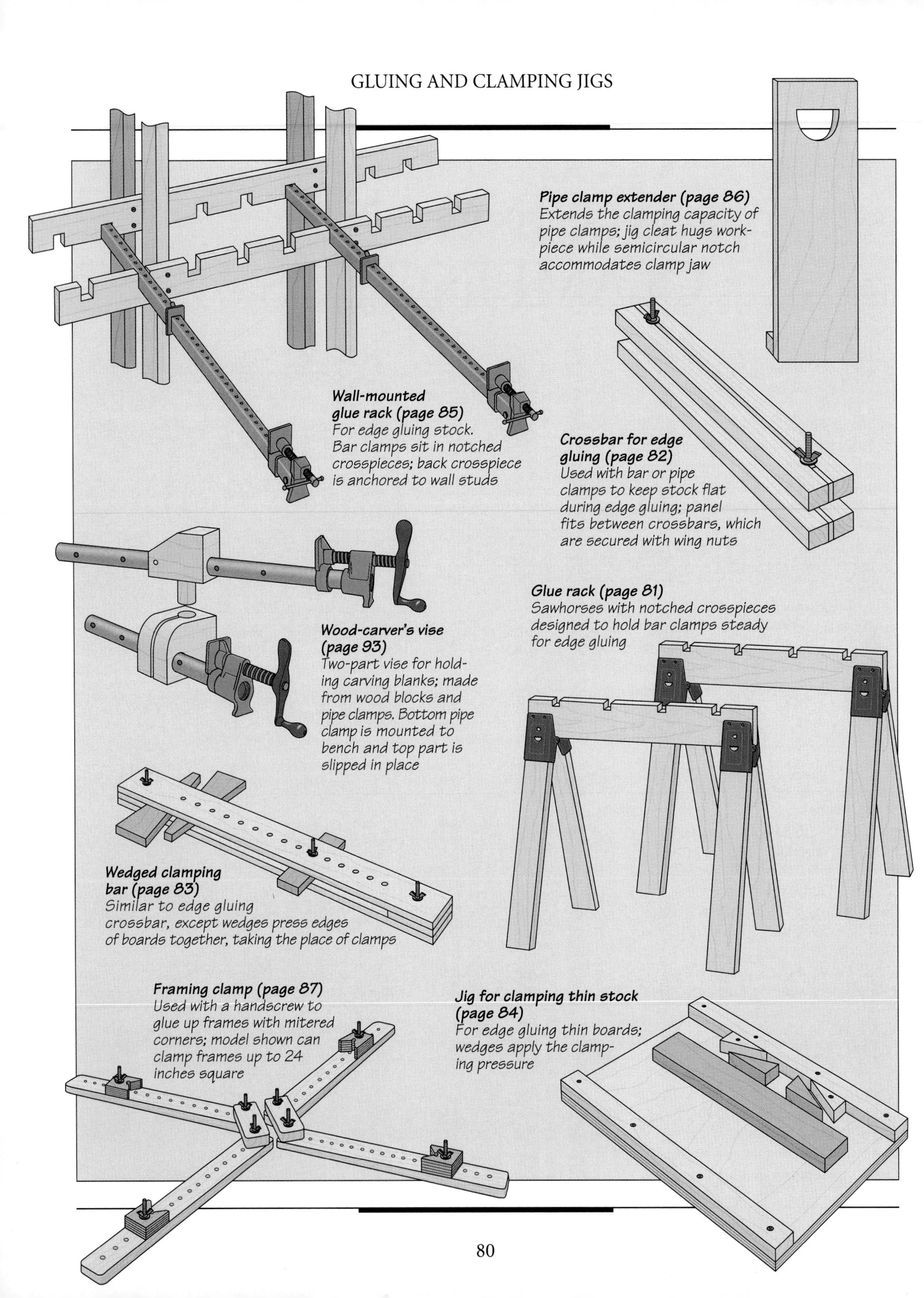
Pipe clamp extender (page 86)
Extends the clamping capacity of pipe clamps; jig cleat hugs workpiece while semicircular notch accommodates clamp jaw
Wall-mounted glue rack (page 85)
For edge gluing stock. Bar clamps sit in notched crosspieces; back crosspiece is anchored to wall studs
Crossbar for edge gluing (page 82)
Used with bar or pipe clamps to keep stock flat during edge gluing; panel fits between crossbars, which are secured with wing nuts
Glue rack (page 81)
Sawhorses with notched crosspieces designed to hold bar clamps steady for edge gluing
Wood-carver's vise (page 93)
Two-part vise for holding carving blanks; made from wood blocks and pipe clamps. Bottom pipe clamp is mounted to bench and top part is slipped in place
Wedged clamping bar (page 83)
Similar to edge gluing crossbar, except wedges press edges of boards together, taking the place of clamps
Framing clamp (page 87)
Used with a handscrew to glue up frames with mitered corners; model shown can clamp frames up to 24 inches square
Jig for clamping thin stock (page 84)
For edge gluing thin boards; wedges apply the clamping pressure

GLUE RACK

1 Building the jig
A pair of racks made from two sawhorses like the one shown at right provides a convenient way to hold bar clamps for gluing up a panel. Remove the crosspiece from your sawhorses and cut replacements the same width and thickness as the originals, making them at least as long as the boards you will be assembling. Cut notches along one edge of each crosspiece at 6-inch intervals, making them wide enough to hold a bar clamp snugly and deep enough to hold the bar level with the top of the crosspiece. You can also cut notches to accommodate pipe clamps, but it is better to use bar clamps with this jig since they will not rotate.

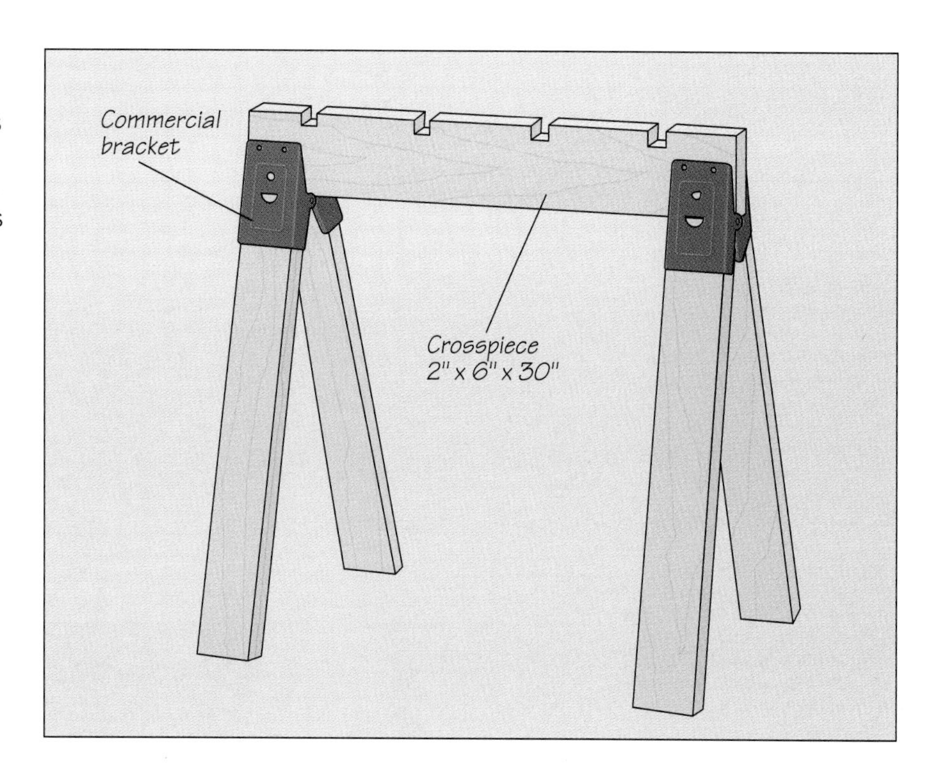

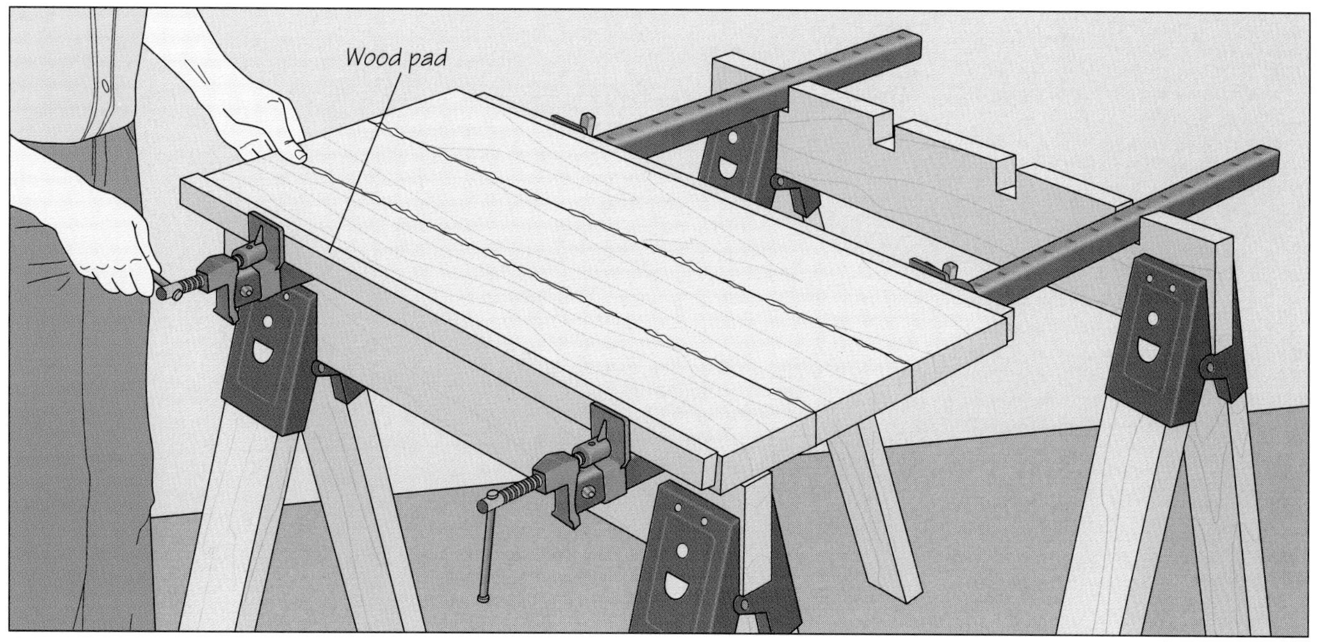

2 Gluing up a panel
Seat at least two bar clamps in the notches so that the boards to be glued are supported every 24 to 36 inches. To avoid marring the edges of the panel when you tighten the clamps, use two wood pads that extend the full length of the boards. Set the boards face-down on the clamps and align their ends. Tighten the clamps just enough to butt the boards together *(above)*, then place a third clamp across the top of the boards, centering it between the others. Finish tightening all the clamps until there are no gaps between the boards and a thin bead of adhesive squeezes from the joints.

CROSSBARS FOR EDGE GLUING

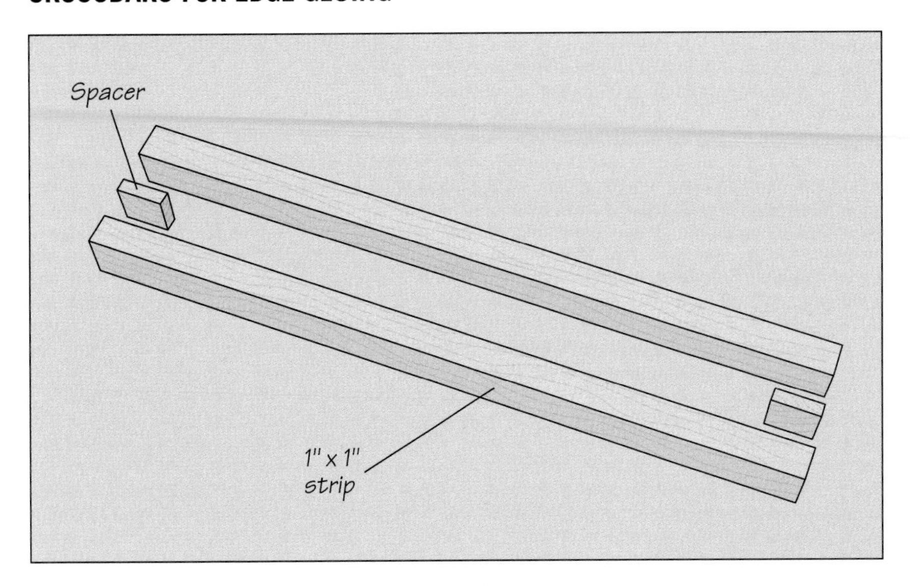

1 Building the crossbars
To keep panels from bowing during glue-up when clamping pressure is applied, bolt a pair of crossbars like the one shown at left between each pair of clamps. Make each crossbar from two short wood spacers and two strips of 1-by-1 hardwood stock a few inches longer than the panel's width. The spacers should be slightly thicker than the diameter of the bolts used to hold the crossbars in place. Glue the spacers between the ends of the strips, and spread wax on the crossbars to prevent excess glue from adhering to them.

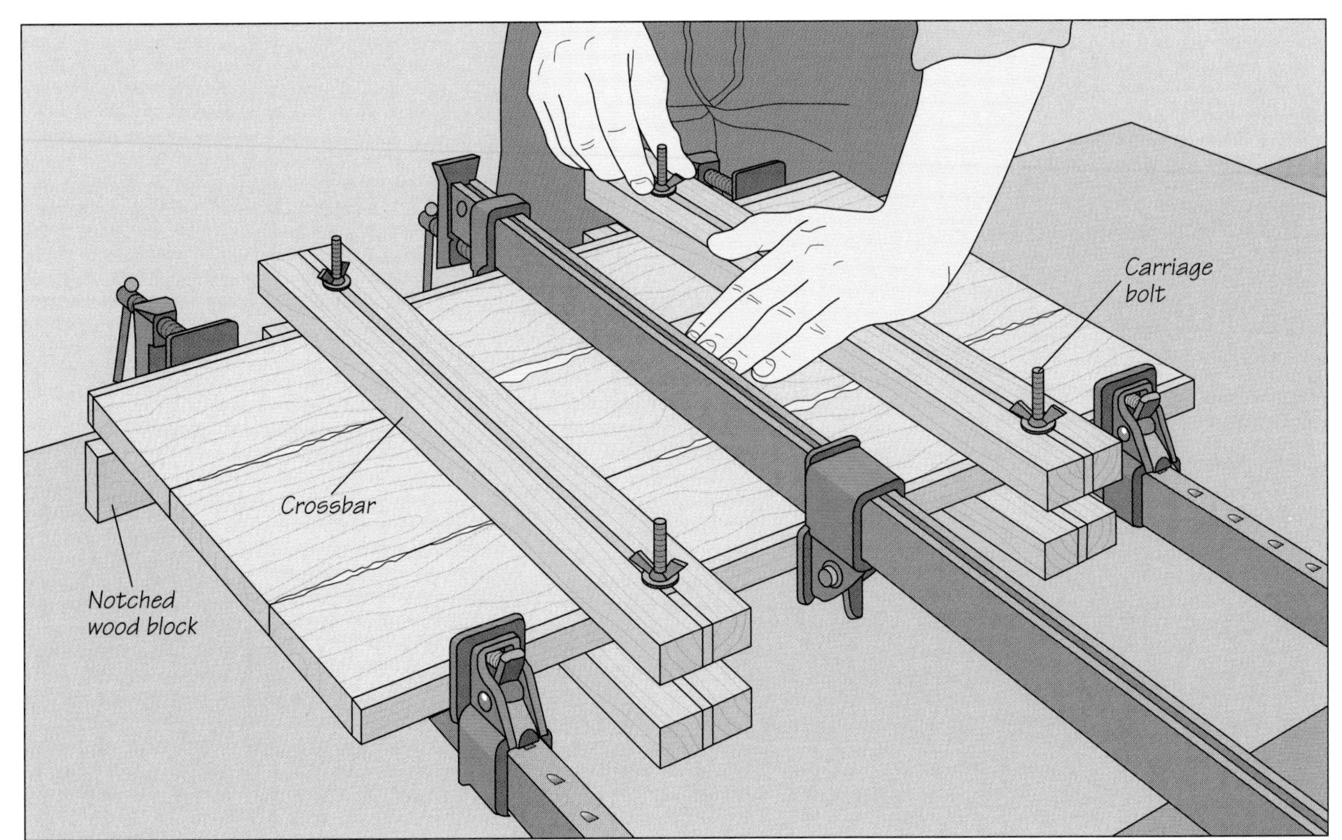

2 Installing the crossbars
Glue up the boards as you would on a rack *(page 81)*. To prevent the bar clamps from tipping over, place the end of each one in a notched block of wood. Before the bar clamps have been fully tightened, install the crossbars in pairs, centering them between the clamps already in place. Insert carriage bolts through the crossbar slots, using washers and wing nuts to tighten the jig snug against the panel *(above)*. Then, tighten the bar clamps completely.

WEDGED CLAMPING BAR

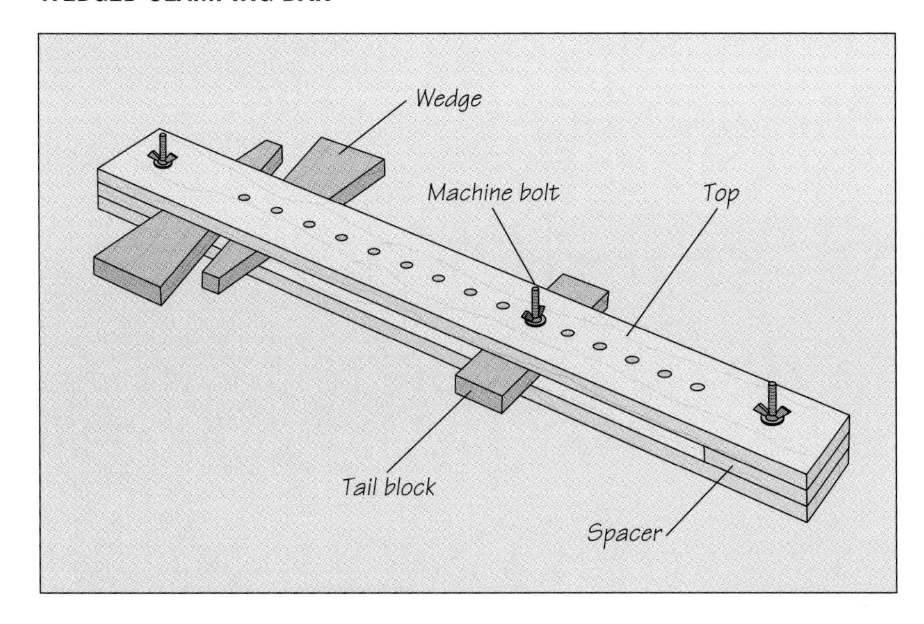

1 Building the jig body
The wedged clamping bar shown at left is an excellent alternative to a bar clamp for edge gluing boards, because it prevents the stock from bowing when pressure is applied. Cut the top and bottom from 3/4-inch-thick stock, making them longer than the widest panel you will glue up. Cut the spacer, tail block, and wedges from stock the same thickness as the boards to be glued. (Keep sets of spacers, tail blocks, and wedges on hand to accommodate boards of varying thickness.) Use a machine bolt, washer, and wing nut at each end of the jig to secure the top, bottom, and spacers together. Wax the bars to prevent adhesive from bonding to them.

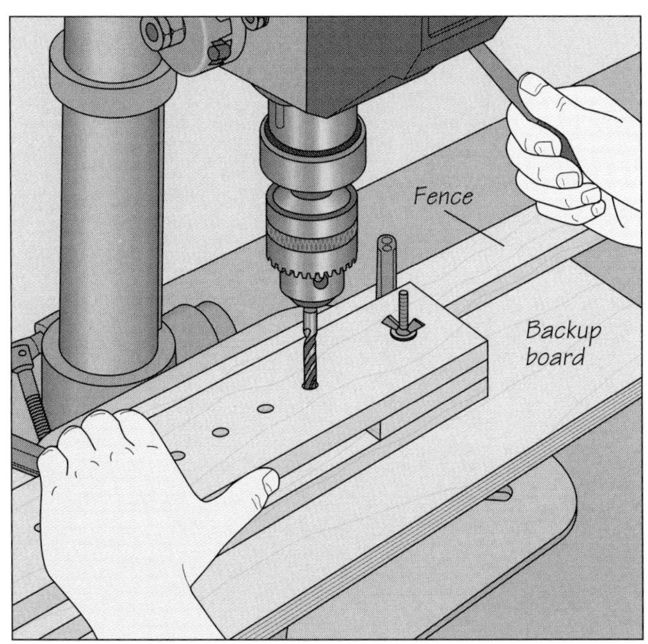

2 Preparing the jig for glueup
You need to bore holes through the jig to adjust it for the width of the panel to be glued. Since you will be drilling straight through the jig, clamp a backup board to your drill press table with a fence along the back edge to ensure the holes are aligned. Install a bit the same diameter as the machine bolt and place the tail block in place. Butt the jig against the fence and drill a hole through the top, the bottom, and the tail block. Bore the remaining holes through the jig body at 1 1/2-inch intervals *(above)*.

3 Edge gluing boards
Spread adhesive on the edges of your stock and set the boards face-down on a work surface. Slip a clamping bar over the boards and position it 6 to 12 inches from one end of the assembly. Butt the tail block against the far edge of the boards, using the machine bolt, washer, and wing nut to fix it in place. To apply clamping pressure, tap one of the wedges at the front edge of the panel *(above)* until there are no gaps between the boards and a thin glue bead squeezes out of the joints. Install the bars at 18- to 24-inch intervals.

A JIG FOR CLAMPING THIN STOCK

Edge gluing thin stock

The benchtop jig shown at right allows you to apply the correct clamping pressure for edge gluing thin stock. Cut the base from ½-inch plywood and the remaining pieces from solid stock. Refer to the illustration for suggested dimensions, but be sure the base is longer than the boards to be glued and the spacer is long enough to butt against the entire front edge of the panel. The edging strips should be thicker than your panel stock. Screw them along the edges of the base and fasten two wedges flush against one strip with their angled edges facing as shown at right. Wax the top face of the base to keep the panel from adhering to it. Apply glue to your stock and set the pieces on the base, butting the first board against the edging strip opposite the wedges. Butt the spacer against the last board and slide the two loose wedges between the spacer and the fixed wedges. Tap the wedges tight to apply clamping pressure *(below)*.

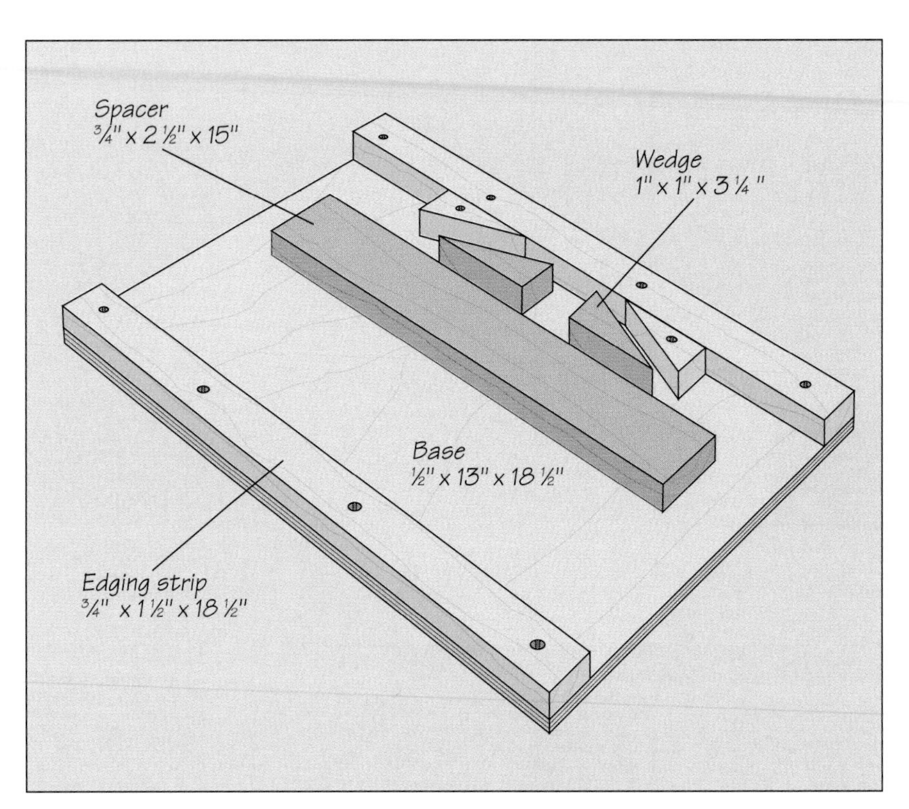

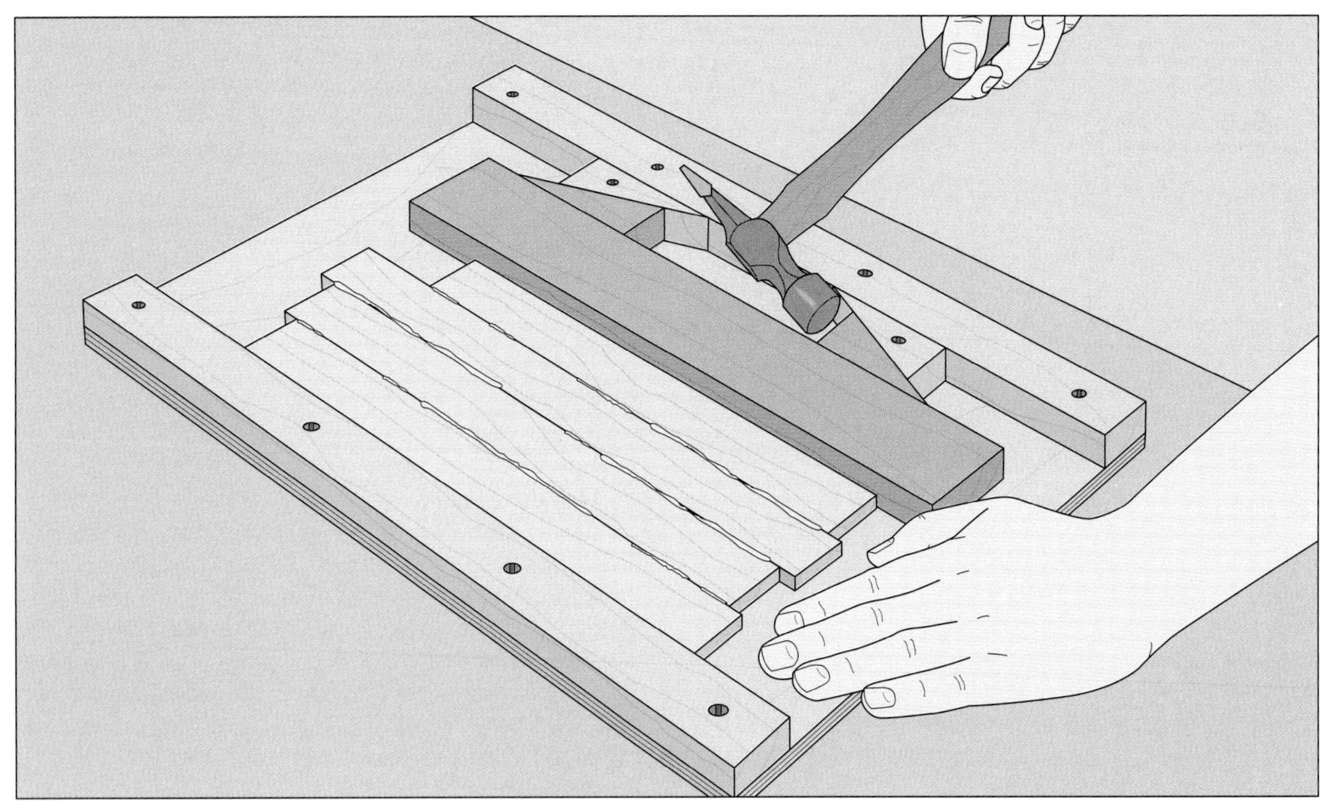

A WALL-MOUNTED GLUE RACK

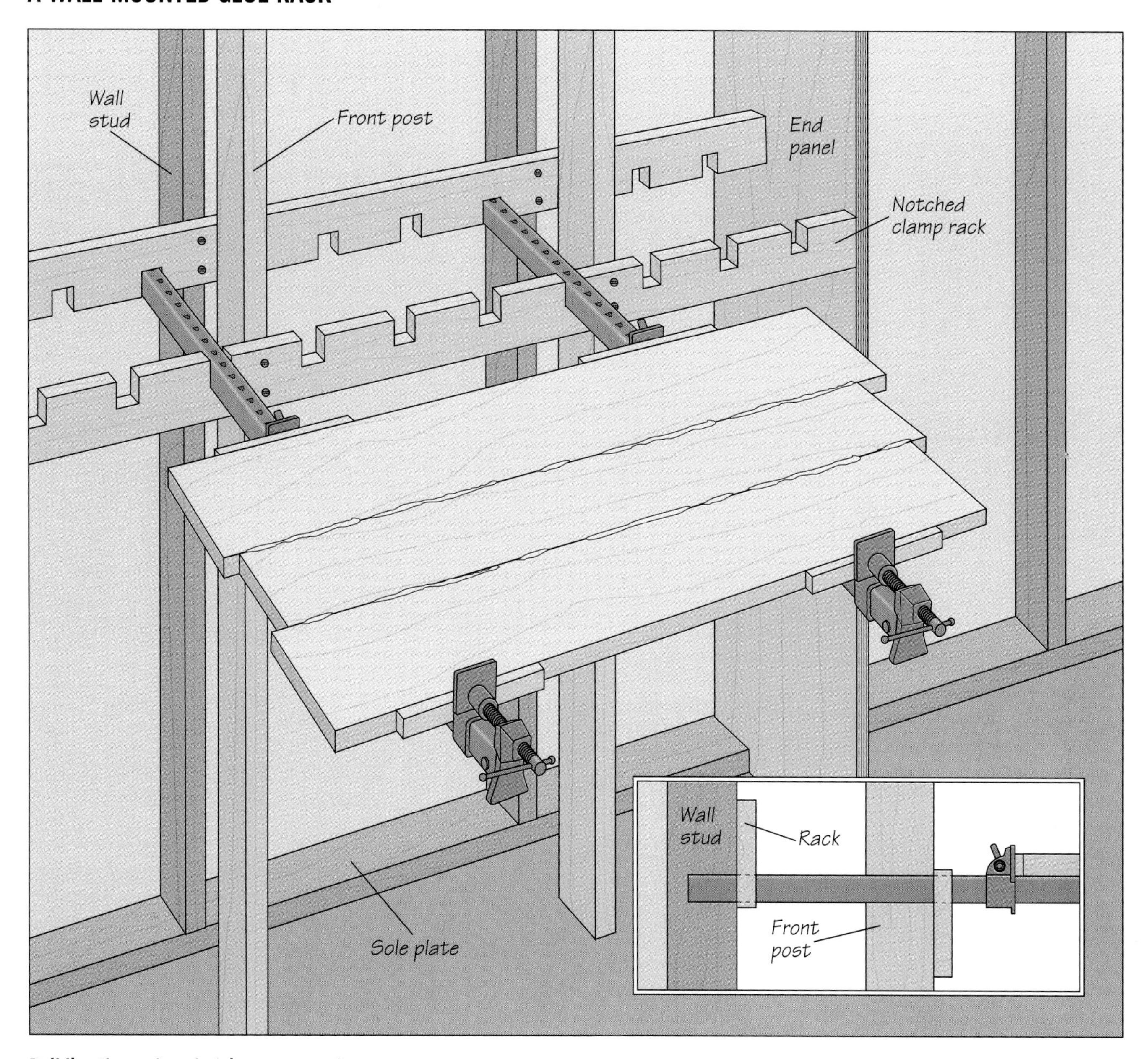

Building the rack and gluing up a panel

The jig shown above allows you to glue up panels using bar clamps, but saves shop space by being mounted to a wall. For clarity, the illustration shows only one pair of clamp racks, but you can install as many as you like from floor to ceiling at 12-inch intervals. Cut the clamp racks from 8-foot-long 1-by-4s and saw notches along one edge of each piece as you would for a sawhorse rack *(page 81)*. Attach one rack of each pair to the wall, driving two screws into every wall stud; make sure the notches are pointing down. To support the front clamp rack, cut floor-to-ceiling 2-by-4s as posts and position one directly facing each stud about 8 to 10 inches from the wall. Screw the front rack to these posts, positioning the notches face-up so they will hold the clamps level. Next, mount two ¾-inch plywood end panels to fit around the jig. Notch the bottom end of the panels to fit over the sole plate and fasten the top to the ceiling. Drive screws through the sides of the end panels into the ends of the racks. To use the jig to glue up a panel, slide bar clamps through the notches in the front and back racks, making sure the ends of the clamps extend beyond the stud-mounted rack *(inset)*. The rest of the operation is identical to edge gluing with any other clamp rack.

A PIPE CLAMP EXTENDER

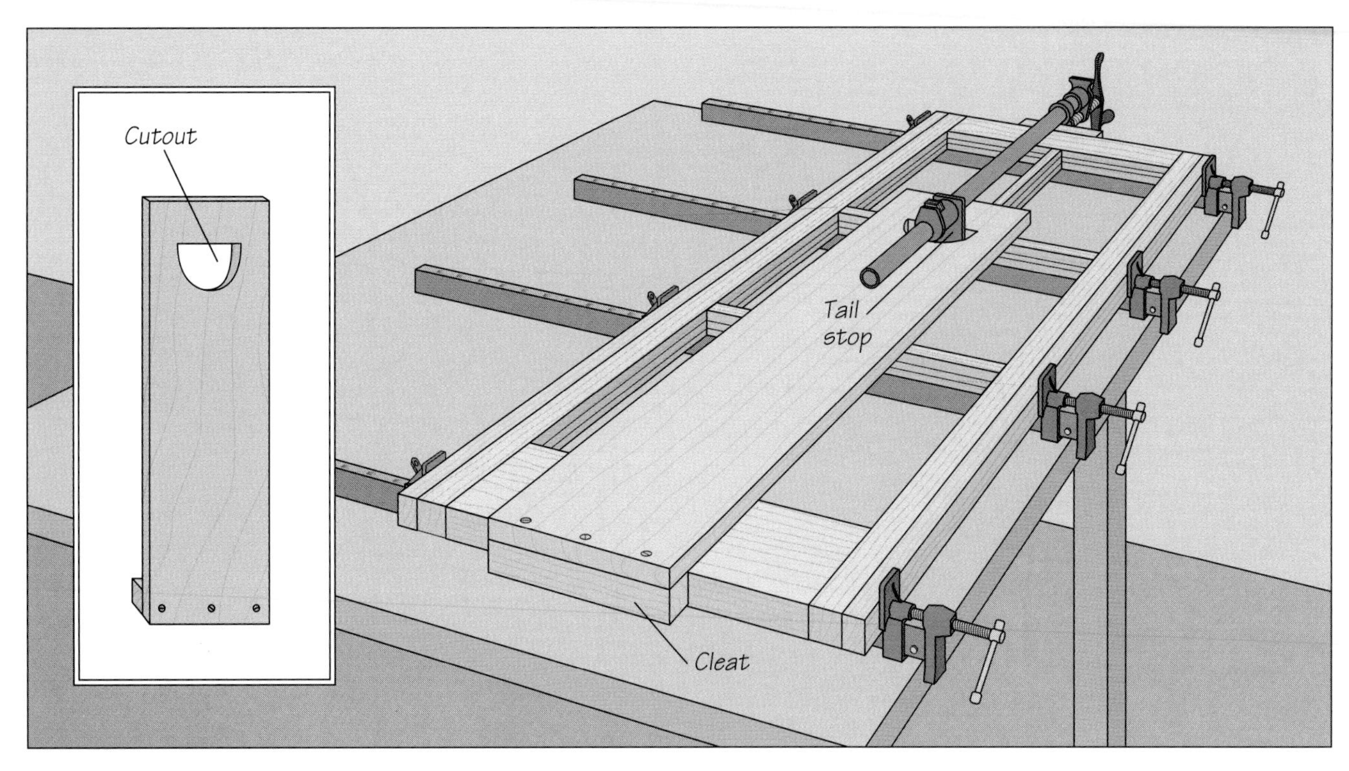

Gluing up a large frame

The jig shown above will extend the capacity of your pipe clamps. Cut the main body of the extender from 1-by-6 stock and the cleat from a 2-by-2. Saw a D-shaped cutout near one end of the body to accommodate the pipe clamp tail stop, then screw the cleat to the opposite end *(inset)*. To apply clamping pressure on the top and bottom rails of a long frame like the one in the illustration, set the cleat against one end of the workpiece and fit the pipe clamp tail stop into the cutout. Then tighten the clamp so that the handle-end jaw is pressing against the opposite end of the workpiece *(above)*. Use wood pads to protect the workpieces.

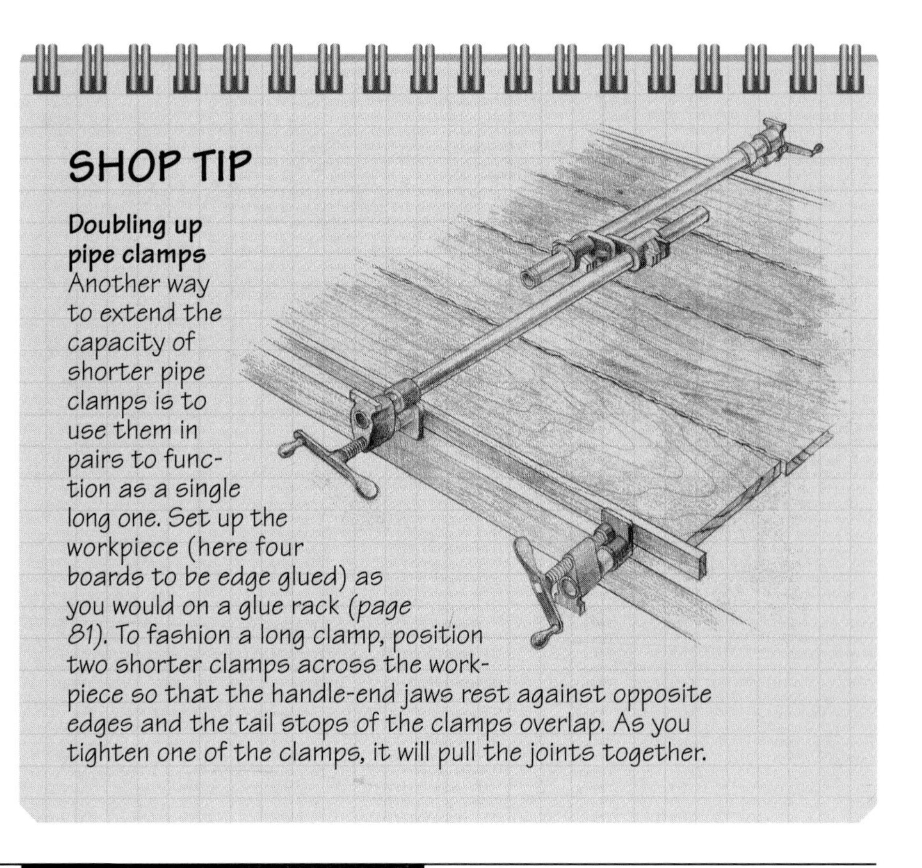

SHOP TIP

Doubling up pipe clamps

Another way to extend the capacity of shorter pipe clamps is to use them in pairs to function as a single long one. Set up the workpiece (here four boards to be edge glued) as you would on a glue rack *(page 81)*. To fashion a long clamp, position two shorter clamps across the workpiece so that the handle-end jaws rest against opposite edges and the tail stops of the clamps overlap. As you tighten one of the clamps, it will pull the joints together.

A FRAMING CLAMP

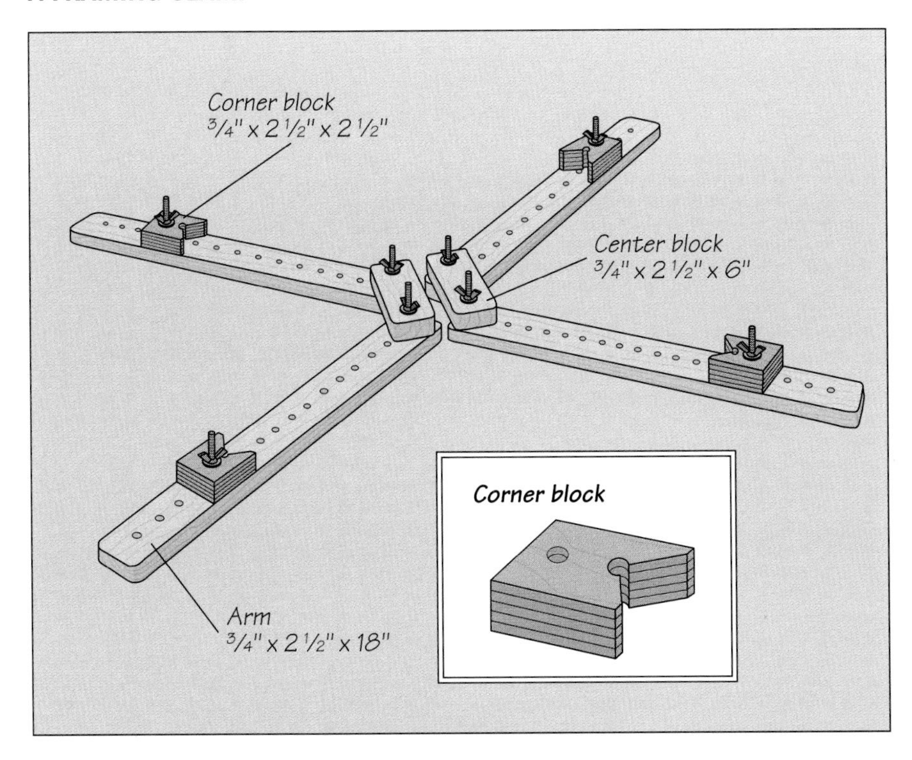

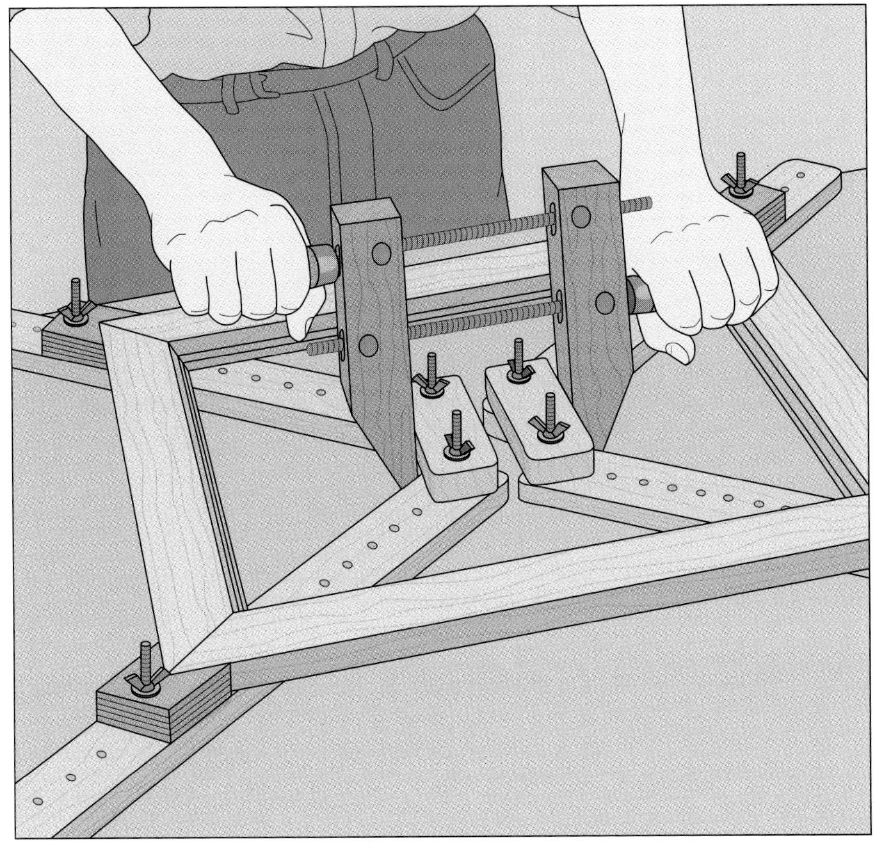

Clamping a mitered picture frame

In conjunction with a handscrew, the self-aligning jig at left is ideal for gluing up frames with mitered corners. The dimensions suggested in the illustration will accommodate frames measuring up to 24 inches on a side. Cut the arms and center blocks from 1-by-3 stock and the corner blocks from ¾-inch plywood. Drill a series of holes down the middle of the arms for ¼-inch-diameter machine bolts; begin 1 inch from one end and space the holes at 1-inch intervals, counterboring the underside to house the bolt heads. Also drill holes through the center blocks 1 inch from each end. To prepare the corner blocks, drill two holes through each one: the first for a machine bolt about 1 inch from one end, and a smaller hole about 1½ inches in from the same end. Finish by cutting a 90° wedge out of the opposite end, locating the apex of the angle at the center of the second hole drilled *(inset)*. Assemble the jig by securing one center block to each pair of arms with bolts, washers, and wing nuts; leave the nuts loose enough to allow the arms to pivot. To clamp a frame, set the jig on a work surface. Fasten the corner blocks to the arms so that the center blocks are about ½ inch apart when the frame lies snugly within the jig. Use a handscrew to pull the center blocks together, tightening the clamp until all the corner joints are closed *(left, below)*.

MITER CLAMPING BLOCKS

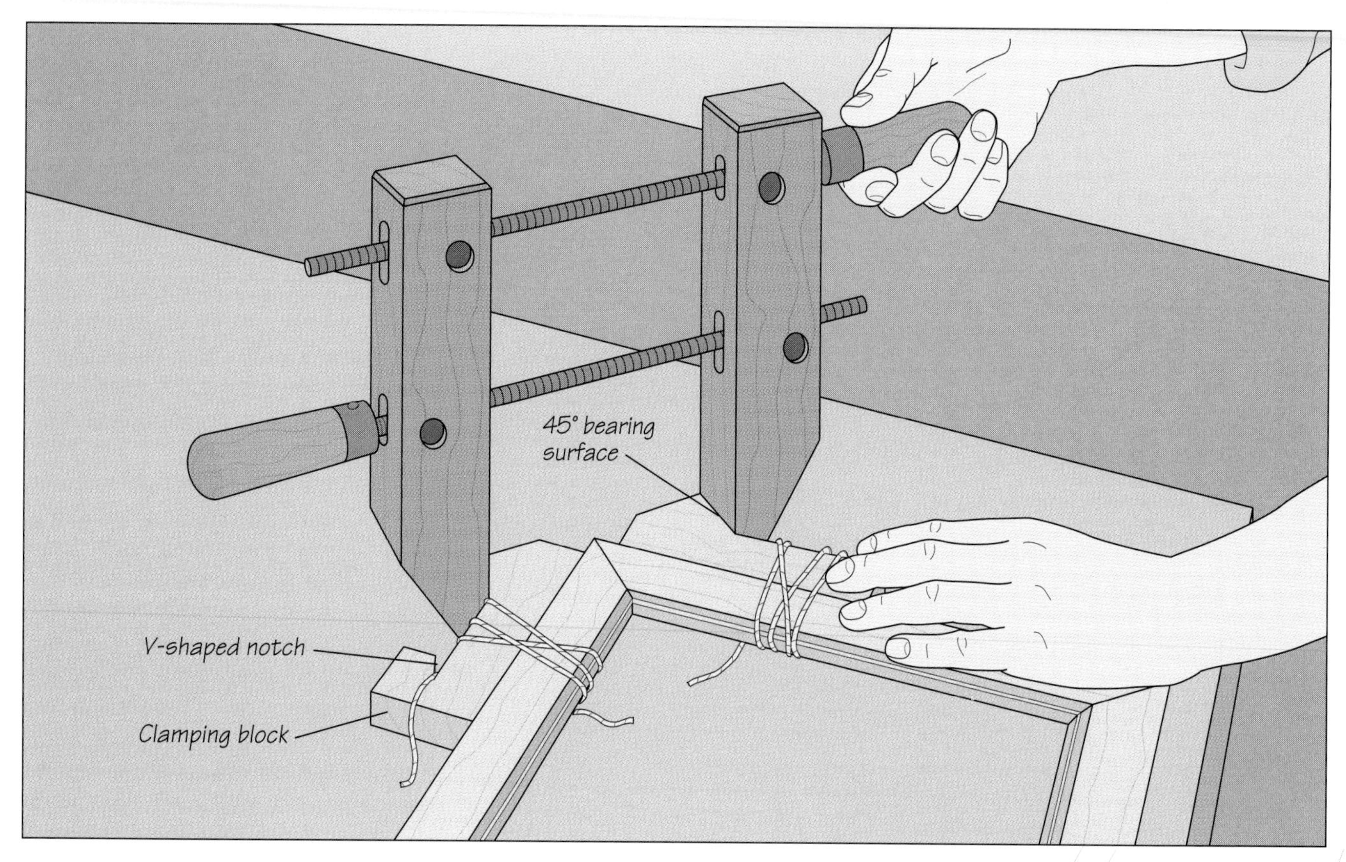

Gluing up a mitered frame

Clamp mitered corners using handscrews and the special blocks shown above. You will need one clamp and two blocks for each corner. Cut the blocks from stock the same thickness as your workpiece. Referring to the illustration above, shape one edge so there is a 45° bearing surface for the handscrew near one end, and a small V-shaped notch near the other. To glue up a corner, tie the blocks snugly to the edges of the frame with cord, securing the loose end in the notch. Set the jaws of the handscrew against the 45° angles and tighten them *(above)* until there are no gaps in the joint and a thin bead of glue squeezes out of it. To keep the joint square, tighten each handscrew a little at a time, checking the corner with a square.

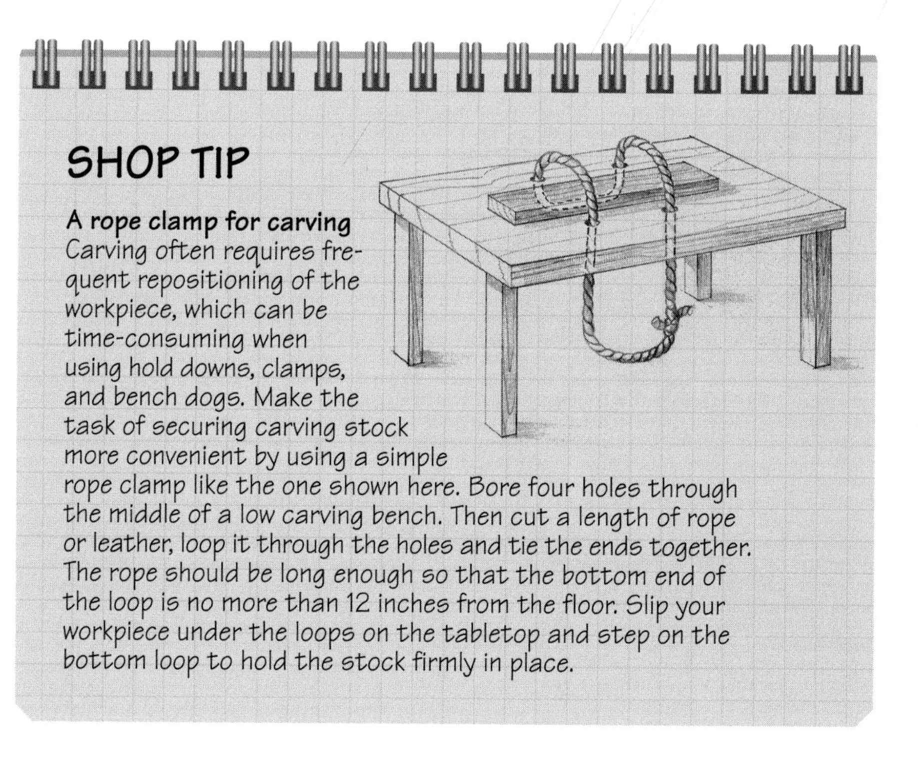

SHOP TIP

A rope clamp for carving

Carving often requires frequent repositioning of the workpiece, which can be time-consuming when using hold downs, clamps, and bench dogs. Make the task of securing carving stock more convenient by using a simple rope clamp like the one shown here. Bore four holes through the middle of a low carving bench. Then cut a length of rope or leather, loop it through the holes and tie the ends together. The rope should be long enough so that the bottom end of the loop is no more than 12 inches from the floor. Slip your workpiece under the loops on the tabletop and step on the bottom loop to hold the stock firmly in place.

A CARCASE-SQUARING BLOCK

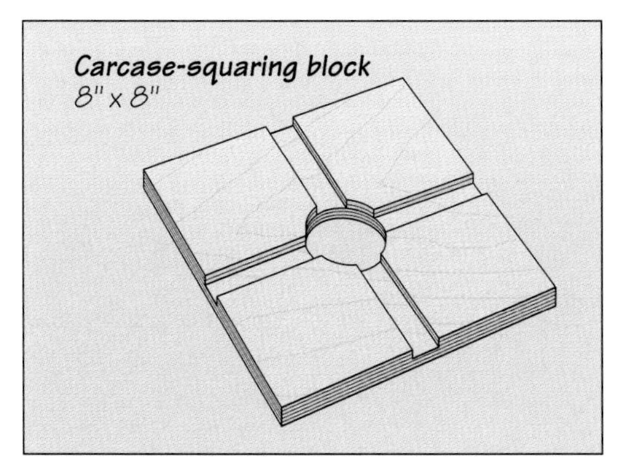

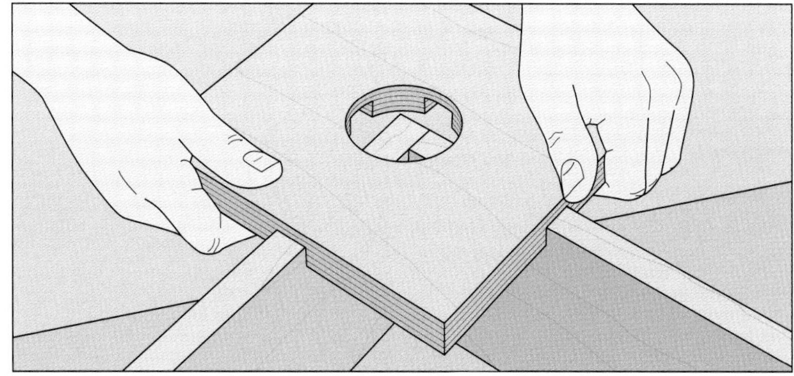

Keeping a carcase square

It can be difficult to keep the four sides of a carcase square during glue-up or while installing a back panel. A carcase-squaring block placed on each corner will solve the problem. Each block consists of an 8-inch square of 3/4-inch plywood. To prevent glue squeeze-out from bonding the block to the carcase, bore a 2-inch-diameter hole in the center of each block with a hole saw or circle cutter. Next, install a dado head on your table saw, adjust it to the same width as the thickness of the carcase stock, and cut two grooves at right angles to one another, intersecting at the center of the block *(above, left)*. To use the jig, apply glue and assemble the carcase, then fit a block over each corner *(above, right)*, centering the hole at the point where the two panels join. Install and tighten the clamps.

TWO WEB CLAMPS

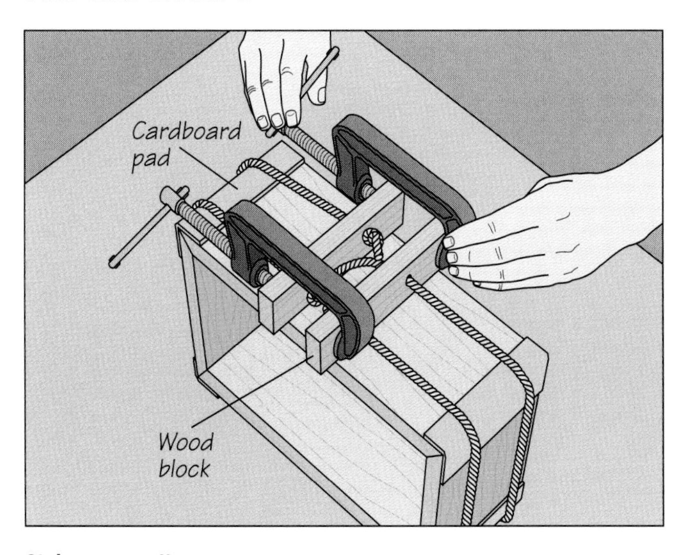

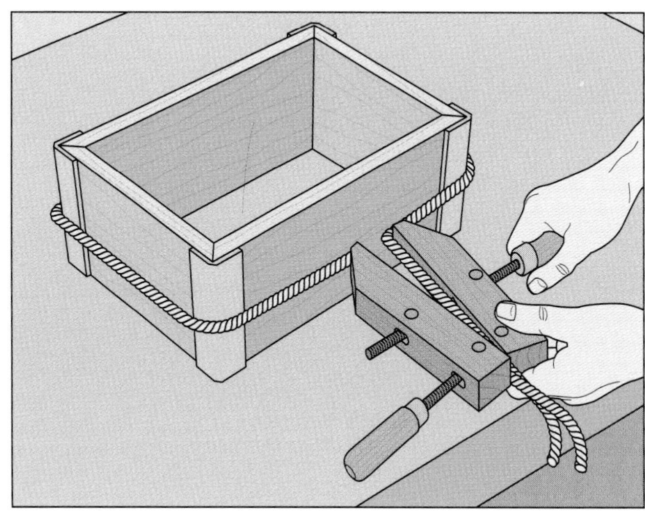

Gluing up small carcases

You can make web clamps out of rope and a handscrew or C clamps. One device uses two wood blocks and two lengths of rope that, when knotted, are slightly shorter than the perimeter of your carcase. Bore two holes through each block near the ends, thread one rope through a hole in each block and knot its ends against the block. Repeat with the other rope, adjusting the length so that the blocks are parallel when set on the carcase. Wrap the ropes and blocks around the carcase, protecting the corners with cardboard pads. Use C clamps to pull the blocks toward each other and clamp the joints *(above, left)*. A second clamping method employs a single handscrew. Wrap a length of rope around the carcase and feed the ends through the clamp. With the tip of the handscrew pressing the rope against the carcase, tighten the back screw to pinch the rope between the back end of the jaws, then close the front end of the jaws to apply clamping pressure *(above, right)*.

You can increase the reach of a C clamp when you need to apply clamping pressure away from the edges of a work surface by using a strip of wood as a clamp extension. To secure the frame shown at right, the strip extends across the frame and clamping pressure is applied over the interior of the frame, securing both sides at once.

A BENCH STOP

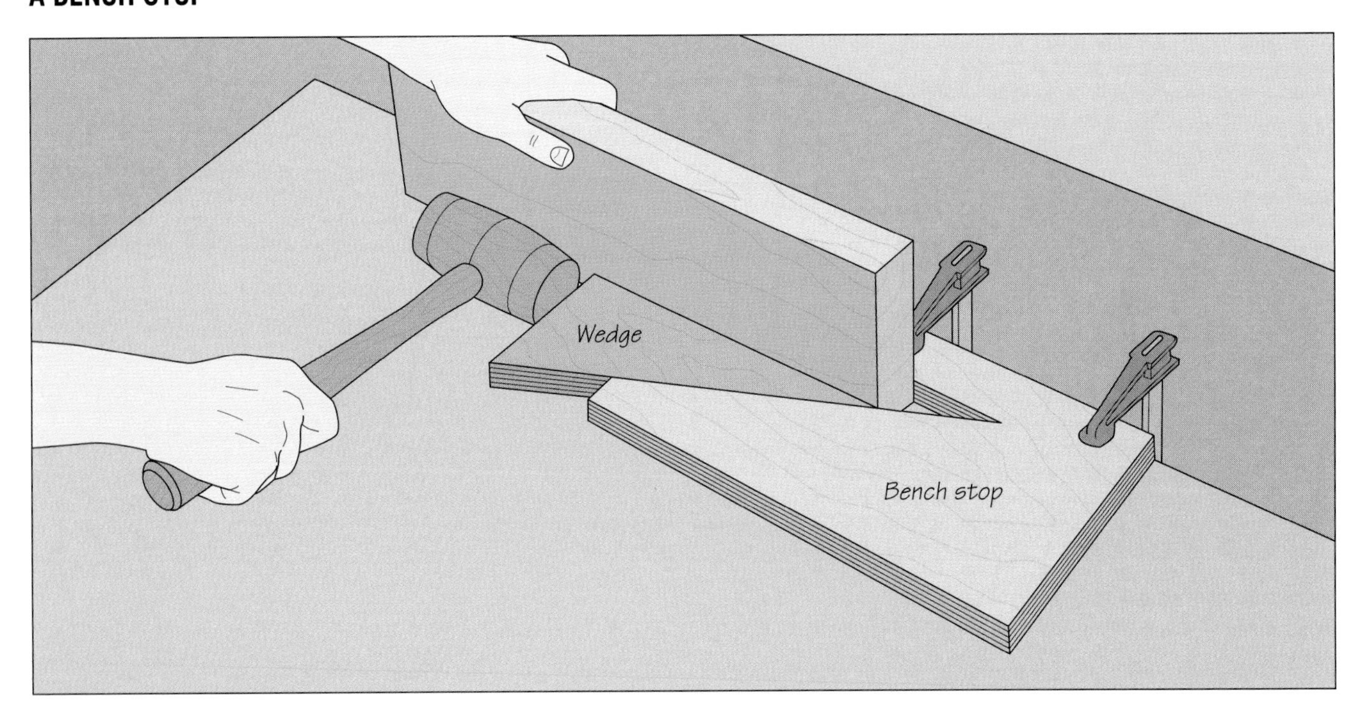

Securing a workpiece to a work surface
You can use a clamped-on bench stop cut from ¾-inch plywood to secure stock to a work surface. Cut the bench stop to size, then mark out a triangular wedge, typically 3 inches shorter than the stop. Saw out the wedge and set it aside. To use the bench stop, clamp it to the work surface and slide the workpiece into the notch, butting one side against the straight edge of the notch. Then tap the wedge tightly in place (above).

BENCH DOGS

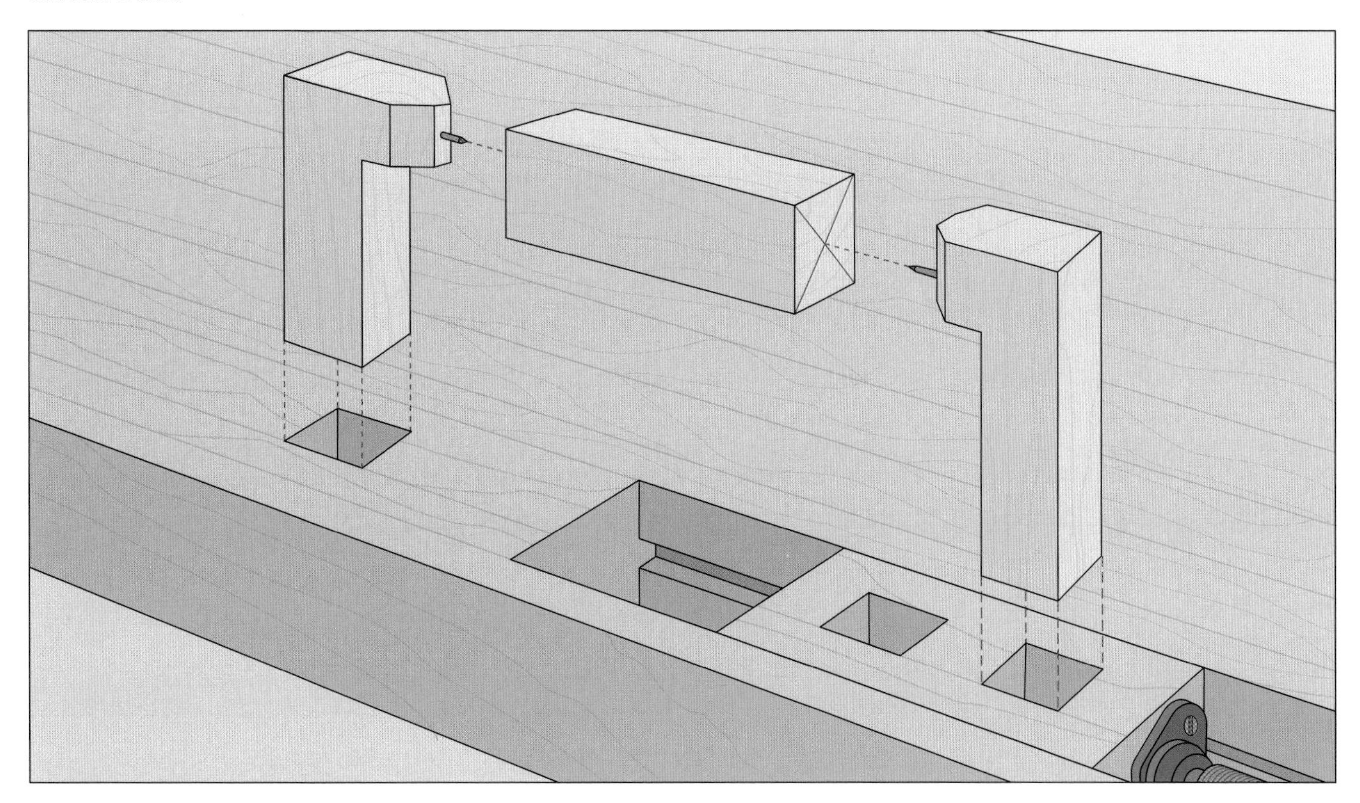

Making and setting up carving dogs

Using a standard bench dog as a model, you can fashion a pair of customized dogs that will grip a carved or turned workpiece, or secure irregular-sized work, such as mitered molding. To make these accessories, cut bevels on either side of the head of a standard bench dog and drive a small screw or nail into the center of the head; snip off the fastener's head to form a sharp point. To use the devices, place one dog in a dog hole of the bench's fixed dog block and the other in the tail vise or a sliding dog block hole *(above)*. Tighten the vise screw until the points contact the ends of the workpiece and hold it securely.

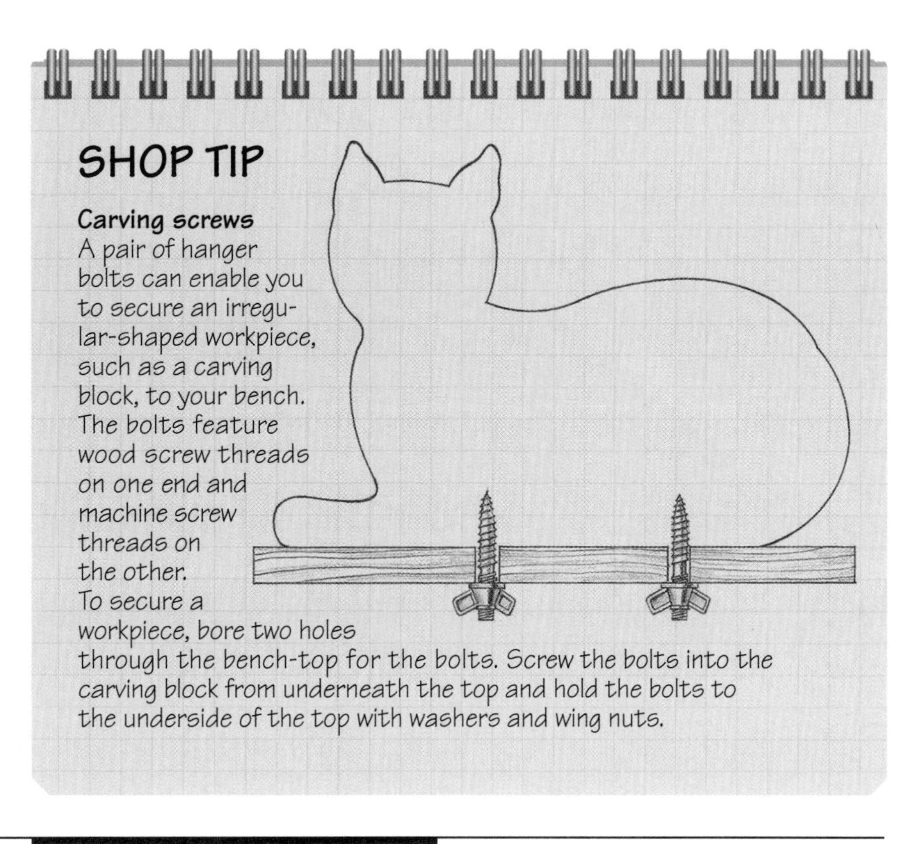

SHOP TIP

Carving screws

A pair of hanger bolts can enable you to secure an irregular-shaped workpiece, such as a carving block, to your bench. The bolts feature wood screw threads on one end and machine screw threads on the other. To secure a workpiece, bore two holes through the bench-top for the bolts. Screw the bolts into the carving block from underneath the top and hold the bolts to the underside of the top with washers and wing nuts.

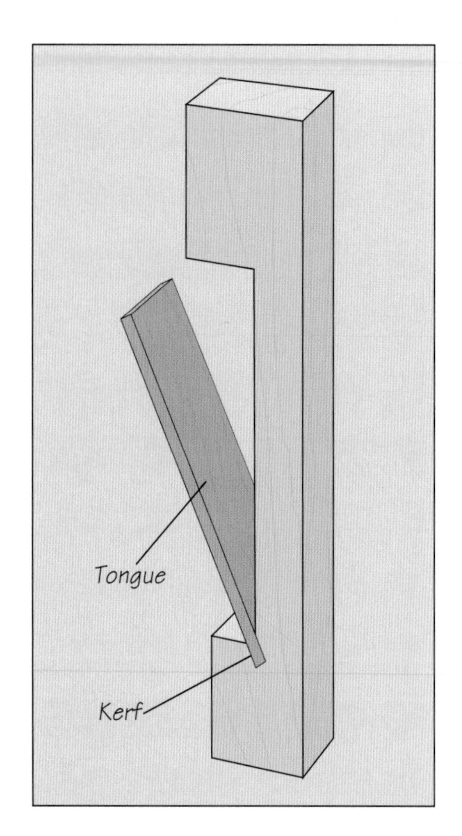

Making a bench dog with a wooden tongue
Bench dogs can be crafted from hardwood stock; the one shown above uses an angled wooden tongue as a spring that presses against the wall of the dog hole to keep the dog in position. Cut the dog to fit the holes in your workbench, then chisel out a notch from the middle of the dog. Saw a short kerf in the lower corner of the notch, angling the cut so the tongue will extend beyond the edge of the notch. Cut the tongue from hardwood, making it about as long as the notch, as wide as the dog, and as thick as the kerf. Glue the tongue in the kerf.

A SUBSTITUTE BENCH VISE

If your workbench does not have a bench vise, you can improvise a substitute using readily available shop accessories. Two large handscrews arranged as shown above will hold a board upright at the corner of the work surface.

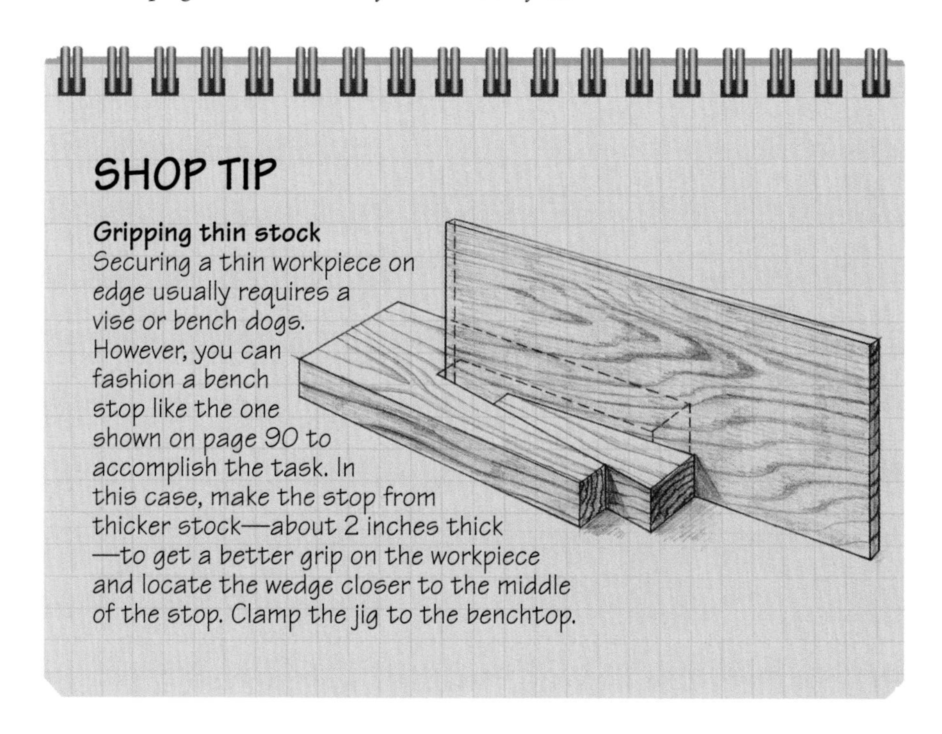

SHOP TIP

Gripping thin stock
Securing a thin workpiece on edge usually requires a vise or bench dogs. However, you can fashion a bench stop like the one shown on page 90 to accomplish the task. In this case, make the stop from thicker stock—about 2 inches thick —to get a better grip on the workpiece and locate the wedge closer to the middle of the stop. Clamp the jig to the benchtop.

A WOOD-CARVER'S VISE

1 Making the jig
To hold carving blanks and curved workpieces at virtually any angle, use an adjustable vise like the one shown at right. Attached to the end of your workbench, the jig consists of two pipe-clamp heads, two lengths of pipe, and two pivoting blocks. The upper block grips the work and swivels horizontally, while the lower block holds the upper block in place and rotates vertically around the pipe. Fashion the two blocks from laminated hardwood stock, referring to the illustration for suggested dimensions. Bore a 1-inch-diameter hole in the upper block's underside and glue in a 2-inch-long dowel, reinforced with a wood screw. Bore a matching hole through the lower block and cut a kerf through the block's rounded edge to the hole. Next, bore a hole into the end of your workbench near one corner, large enough to accommodate a 12- to 14-inch length of pipe. Drill a matching hole through the lower block, positioning it near the rounded edge of the block. Finally, bore a hole for the pipe through the upper block.

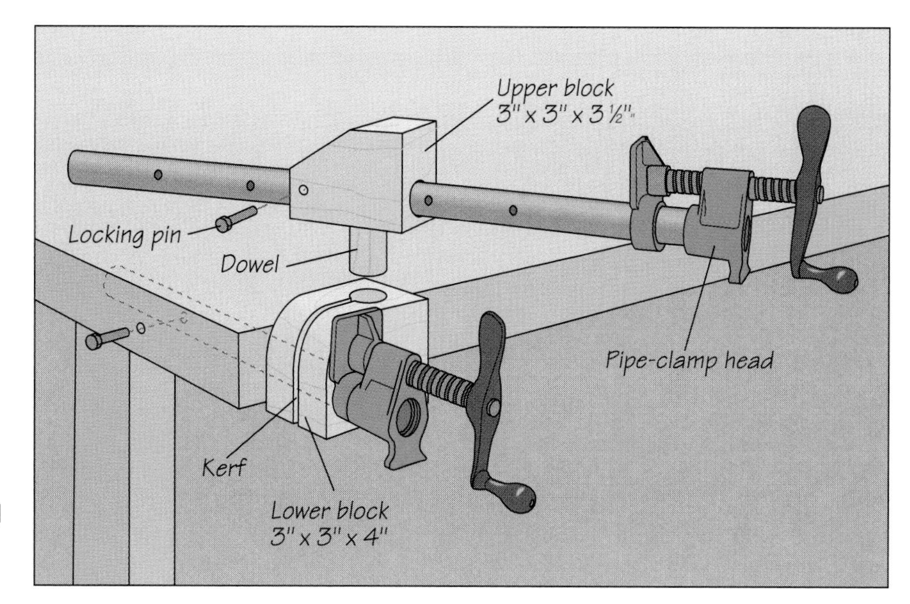

2 Securing a workpiece
Insert the 12- to 14-inch-long pipe with a pipe-clamp head at one end through the bottom block and into the bench. Fix the pipe in place with a locking pin—a small machine bolt or large wood screw. Fit the upper block into the lower one and tighten the lower pipe-clamp head to secure the upper block. Next, insert a longer length of pipe—also with a pipe-clamp head at one end and holes drilled every 4 inches for a locking pin—through the hole in the upper block. Place the workpiece against the upper block and the pipe-clamp's head as shown below, fix the pipe with the locking pin, and tighten the head. To reposition the workpiece, loosen the lower clamp head, pivot the two blocks, and tighten the head.

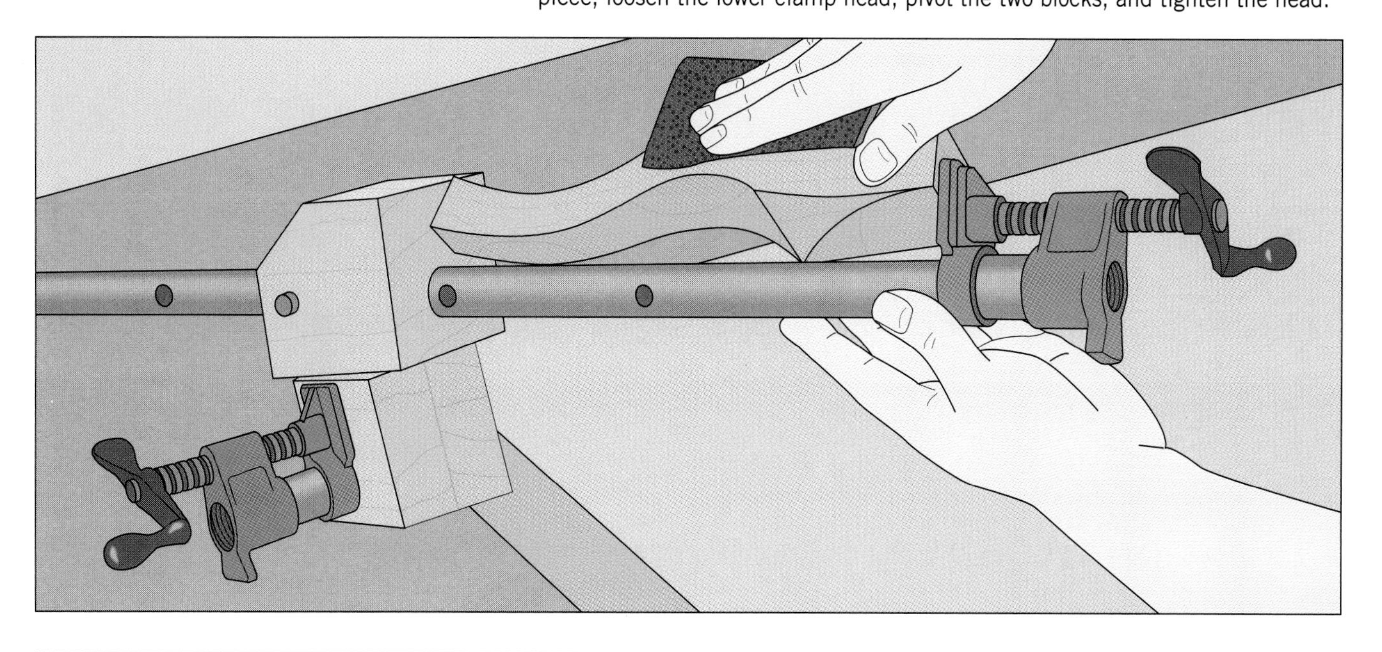

SANDING JIGS

Sanding is one of woodworking's most tedious tasks, but there are tools and techniques to make the job easier and improve your results. The use of a simple jig will help improve the speed and efficiency of most operations, and make tricky tasks easier to accomplish. Several are shown here and on the following pages. A custom-made sanding block *(page 100)*, for example, allows you to smooth contours that would otherwise be difficult to reach. For sanding the edge of circular workpieces, a circle-sanding jig *(page 96)* is designed to hold a belt sander on its side while you feed the stock across the belt. Also useful is a jig for surfacing thin stock *(page 99)* on the radial arm saw.

For everyday sanding tasks, use the auxiliary sanding table shown on page 98 or the sanding block on page 100. The latter device, also shown in the photo at left, offers a simple way of holding a piece of sandpaper in place against a flat surface. To save time and sandpaper, consider the tips on gang sanding and folding sandpaper shown on pages 96 and 101.

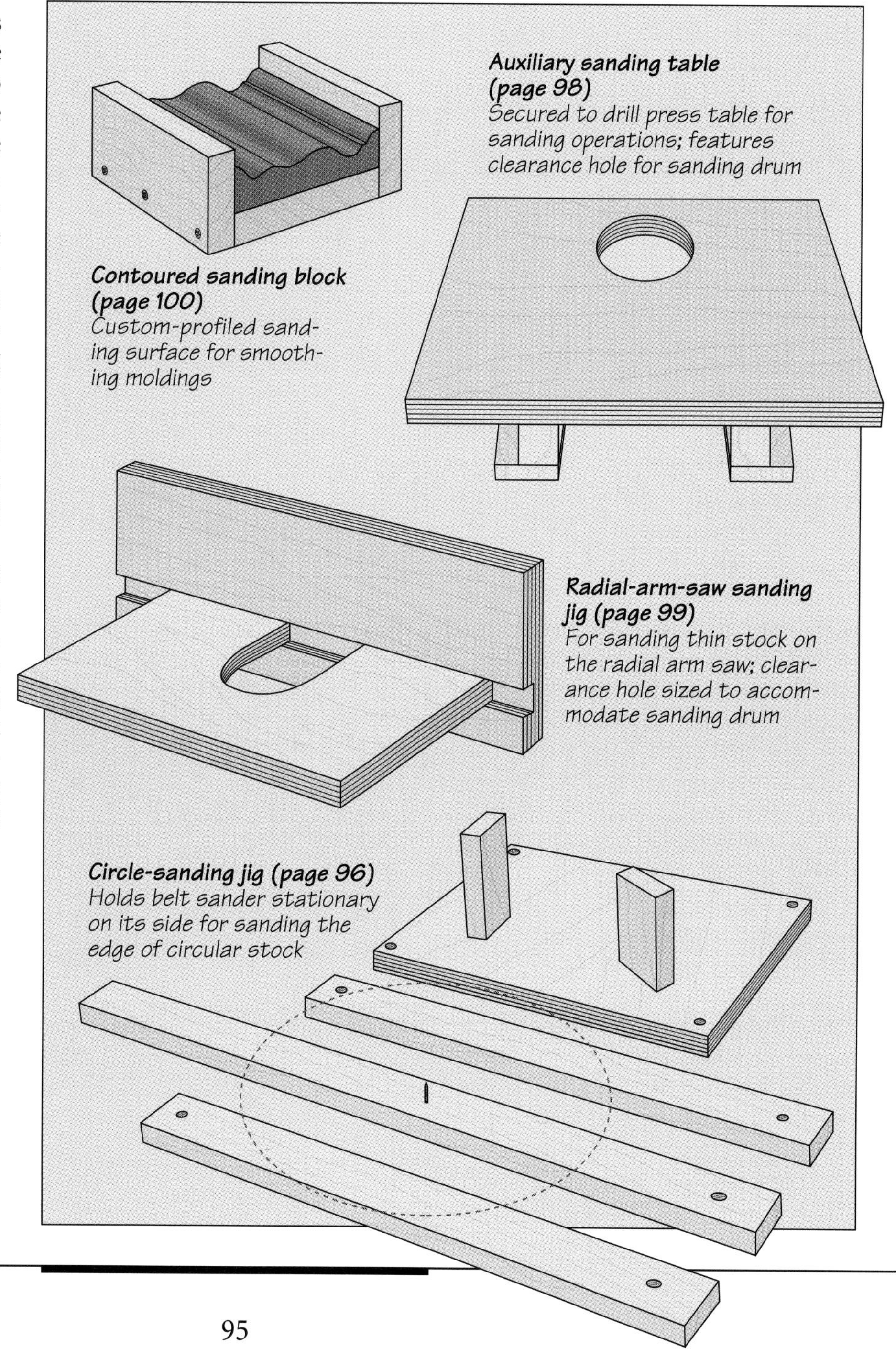

Fitted with 220-grit sandpaper, a shop-made sanding block is used to smooth the surface of the rail joining two table legs. A felt or cork pad can be glued to the bottom face of the block to provide even sanding pressure.

JIGS FOR FACE- AND EDGE-SANDING

You can save time and achieve uniform results by smoothing several workpieces together, a technique known as "gang sanding." If they are the same width, the boards can be secured to a work surface face-to-face and edge-up, and sanded as though they were a single workpiece.

A CIRCLE-SANDING JIG

1 Building the jig
The jig shown at right allows you to smooth the edges of a circular workpiece with a belt sander. The dimensions of the jig will depend on the size of your sander. Make the base and table with 3/4-inch plywood and add solid wood support posts to fit in the handles of the sander. Set the tool on its side on the table and slip the posts in place, then screw them to the table and screw the table to the base. Cut the pivot bar and support boards from 1-by-2 stock, making the boards longer than the diameter of your workpiece and the bar longer than the boards. Screw the support boards to the base so they will support the workpiece near its edge. Next, drive a screw through the middle of the pivot bar and fasten it to the underside of the workpiece. Flip the assembly over and screw one end of the pivot bar to the base midway between the support boards, leaving the screw loose enough to allow you to pivot the bar's other end.

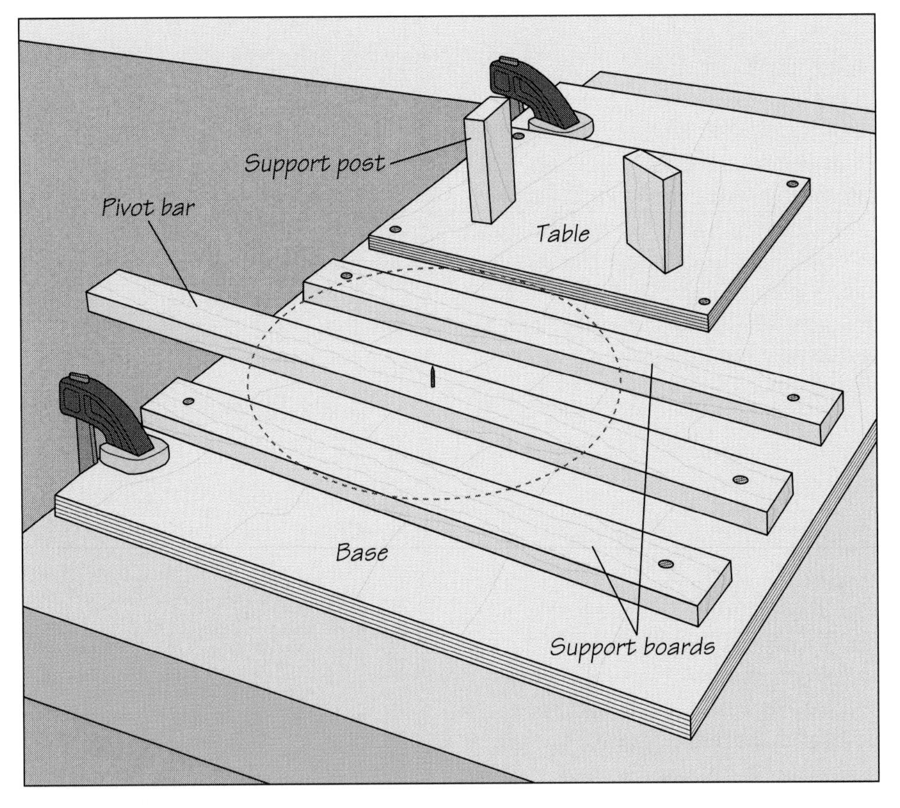

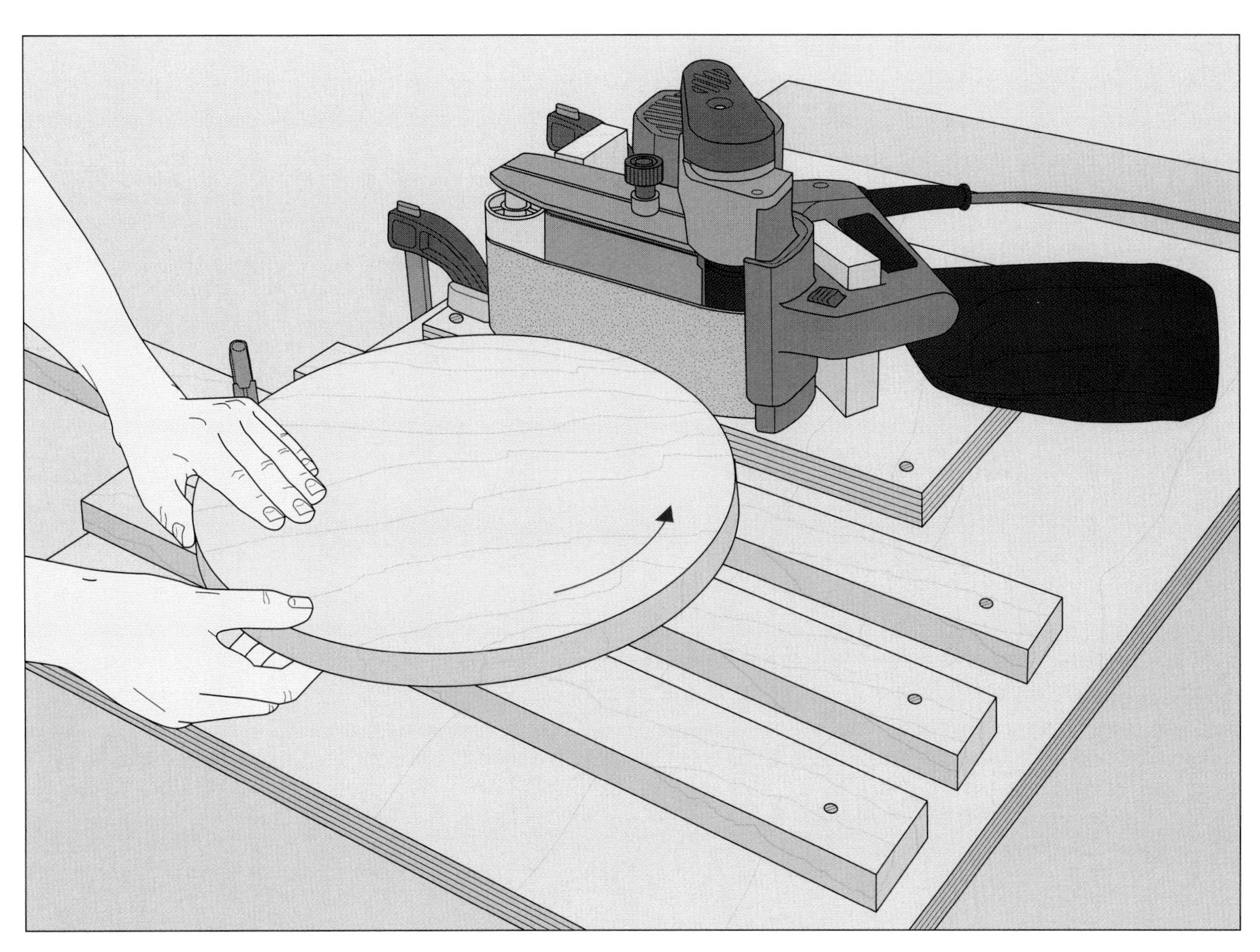

2 Sanding a circular workpiece
Clamp the jig base to a work surface and set the sander in place on the table, fitting the tool's handles over the support posts. Turn on the sander and, if necessary, adjust the tool's tracking mechanism to move the belt down to the jig table. Lock the On/off trigger in the on position and move the free end of the pivot bar toward the sander until the workpiece edge touches the belt. Clamp the free end of the bar to the base, then rotate the piece against the direction of belt rotation *(above)* until the edge is smooth. Keep the workpiece in motion and periodically shift the pivot arm toward the sander to maintain pressure against the sanding belt.

SHOP TIP

Drill press sanding drums

If you need a special off-size sanding drum for your drill press, you can make your own from a dowel. Find the right size dowel, then cut a strip of sandpaper as long as its circumference. Apply a thin coat of glue to the wood and fasten the paper to it *(right, top)*. For a flexible sander that can smooth irregularly shaped workpieces or enlarge holes, cut a slot in a dowel and slip a strip of abrasive paper into it *(right, bottom)*.

A SANDING TABLE FOR THE DRILL PRESS

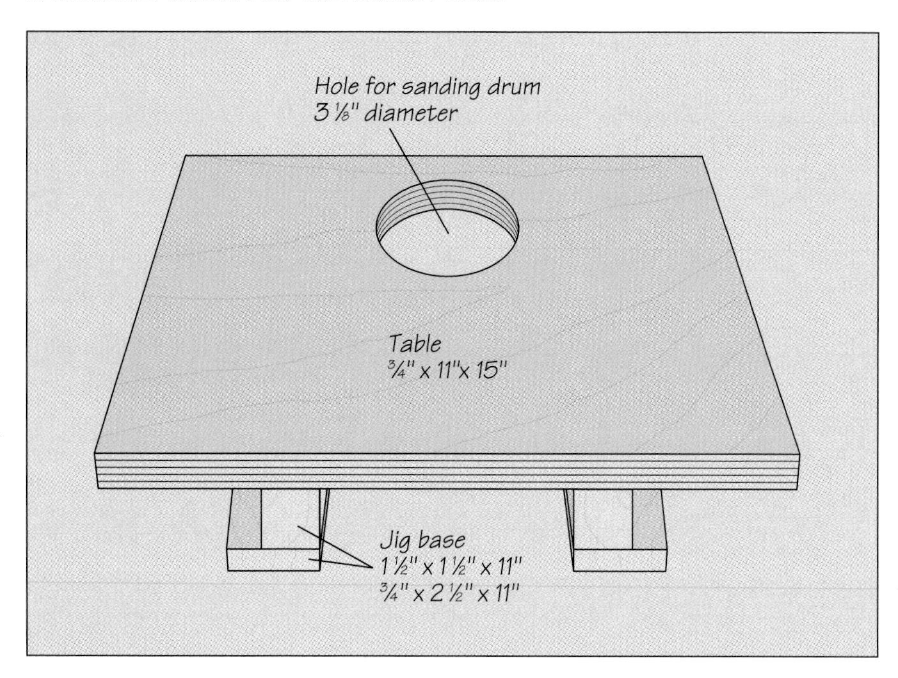

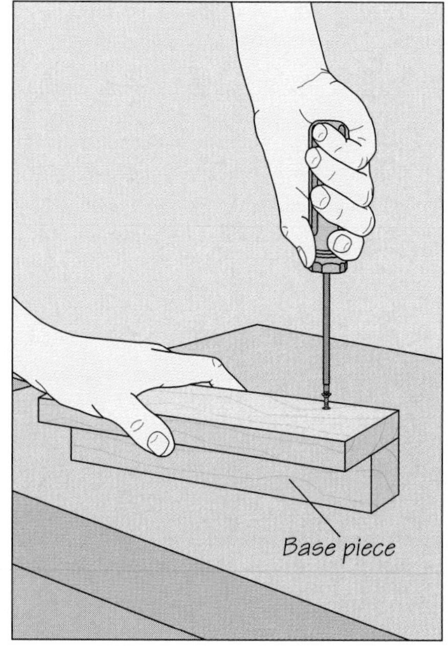

1 Making the jig
Sanding drums larger than ⅞ inch in diameter are too large to fit through the hole in most drill press tables. To make full use of the sanding surface of larger drums, build a table like the one shown above. Use a coping saw, a saber saw, or an electric drill fitted with a hole saw to cut a hole in the plywood top, centering the opening 3 inches from the back of the table. Screw the L-shaped base pieces together from 1-by-3 and 2-by-2 stock *(above, right)*, then glue them to the table.

2 Sanding curved stock
Clamp the jig base to the drill press table with the hole directly underneath the drum. Adjust the table height to bring the bottom of the sanding sleeve level with the jig. Holding the workpiece firmly, feed it at a uniform speed in a direction opposite the rotation of the sanding drum *(right)*. To avoid burning or gouging the workpiece, feed it with a smooth, continuous motion. As segments of the sanding sleeve wear out, raise the drill press table to bring a fresh surface to bear.

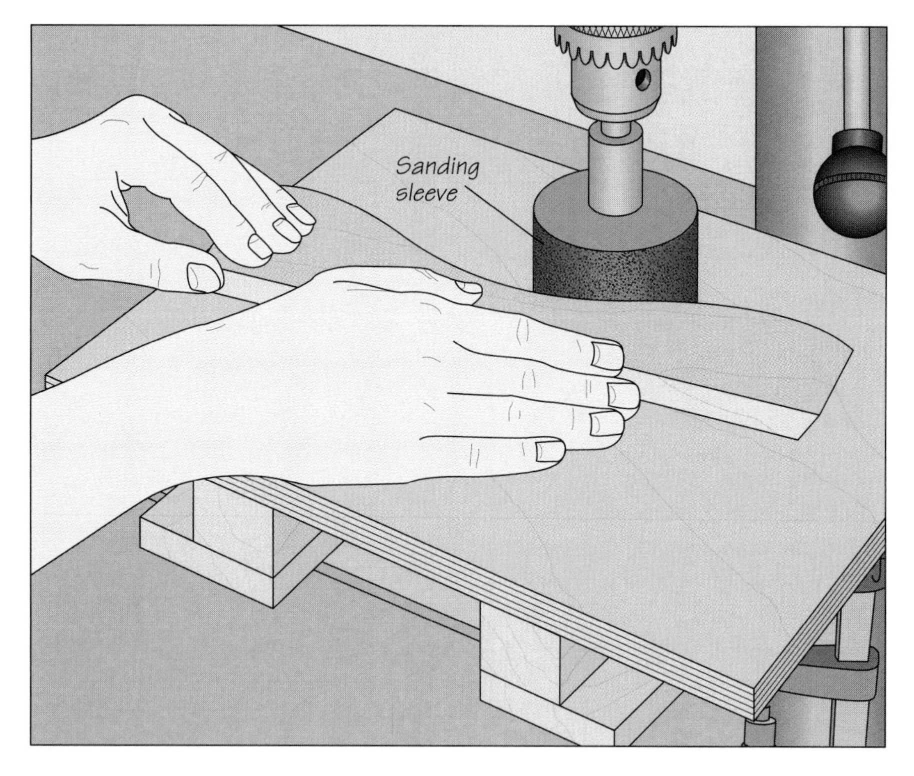

A RADIAL-ARM-SAW SANDING TABLE

1 Building the table
The simple jig shown at right will enable you to sand thin stock on your radial arm saw. Cut the table and fence from ¾-inch plywood, sizing the pieces to fit your saw. Cut a slot out of one edge of the table large enough to accommodate the sanding drum you will use and allow you to shift the drum forward and back for different sized stock. Rout a ¾-inch-wide groove along the length of the fence, leaving ¾ inch of stock below the channel to slip into the saw table's fence slot. Insert the slotted edge of the table into the groove in the fence and screw the two pieces together.

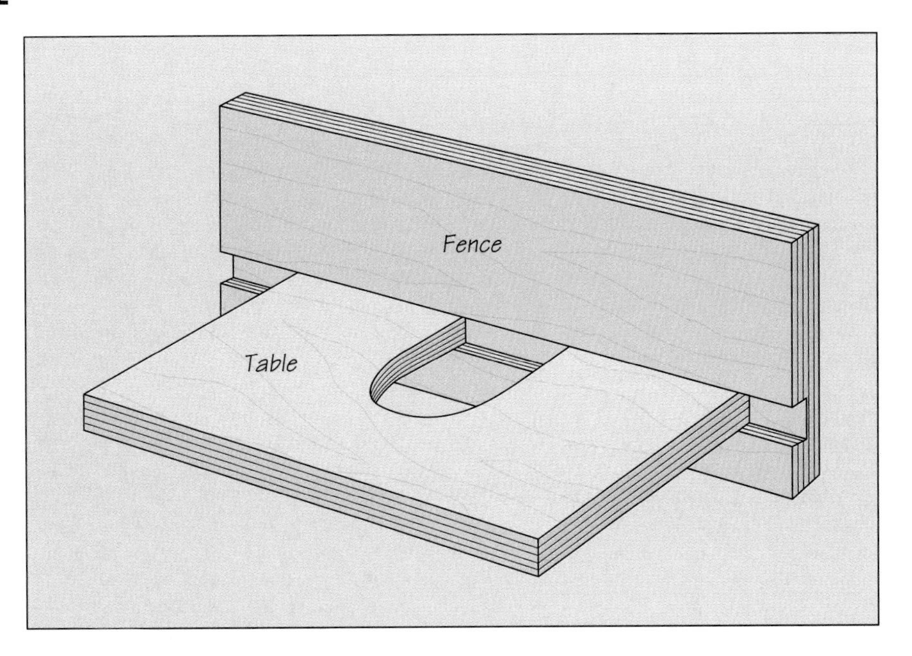

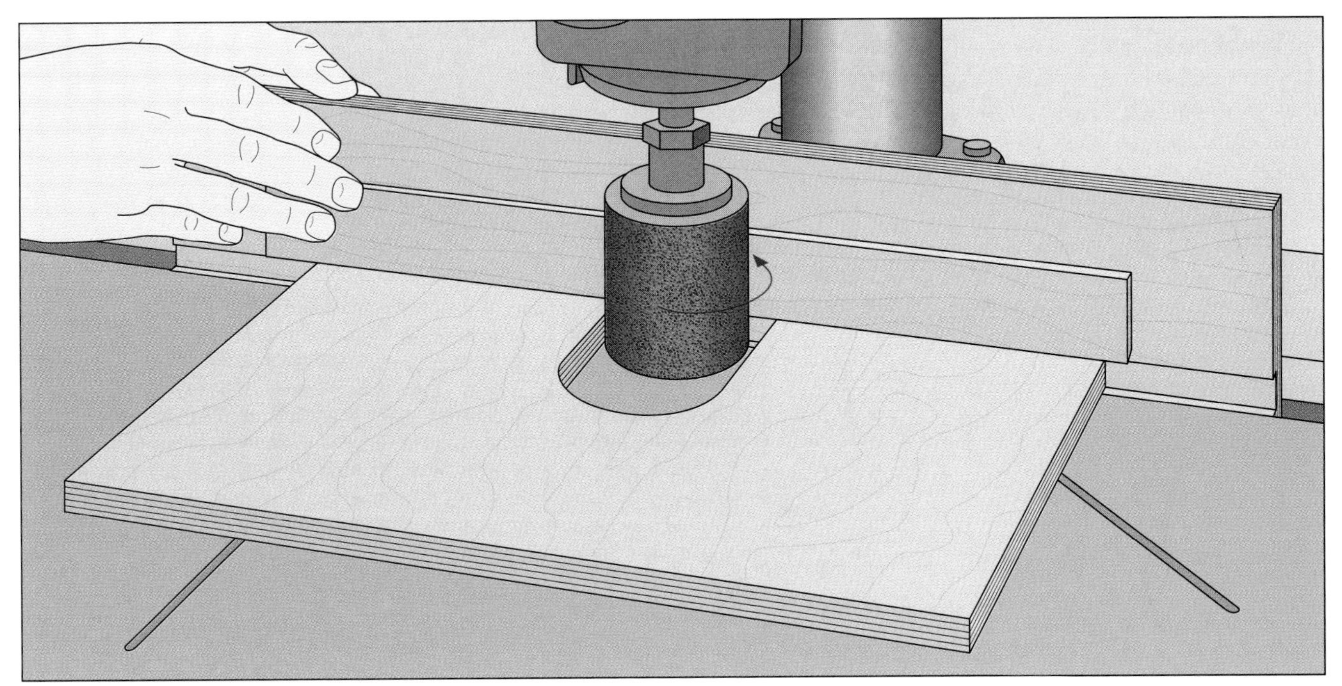

2 Smoothing thin stock
Install a sanding drum in your saw following the manufacturer's instructions. Slip the jig fence between the front and rear tables of the saw, positioning the opening in the base directly below the drum, then tighten the table clamps to secure the jig in place. Lower the drum so it is just below the top of the jig table; position the drum so the distance between it and the jig fence is slightly less than the thickness of the stock you will be sanding. Turn on the saw and feed the stock slowly and continuously from left to right—against the rotation of the drum—between the fence and sanding drum *(above)*. (Clamping a featherboard to the jig table to press the stock against the fence can also help prevent gouging.) For each successive pass, reduce the gap between the fence and sanding drum by no more than 1/16 inch.

A CONTOURED SANDING BLOCK

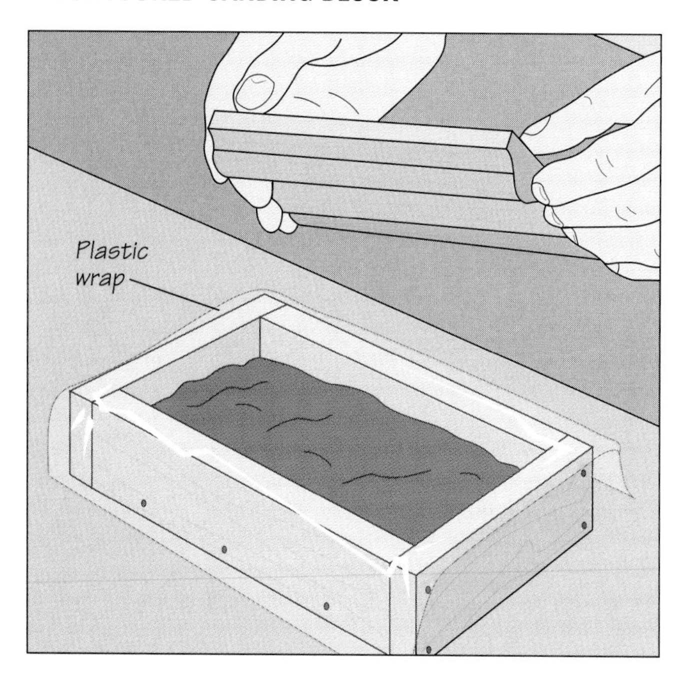

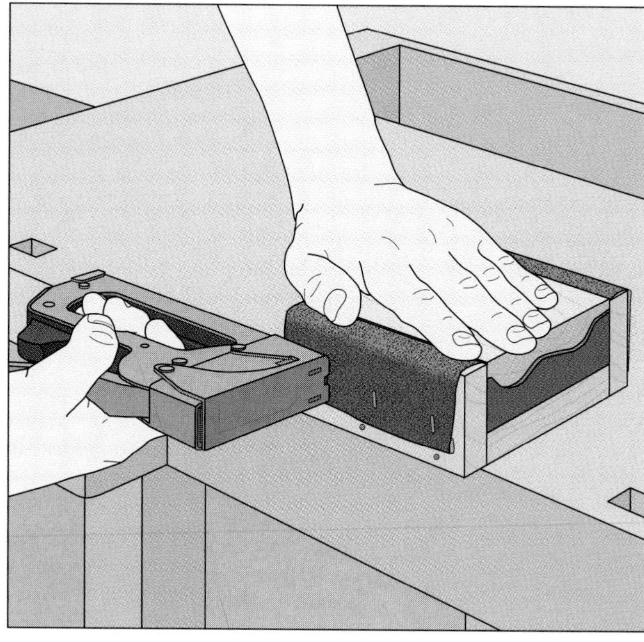

1 Making the block

For smoothing the contours of a piece of molding, you can use a short sample of the molding to shape a sanding block that mates perfectly with the surface of the workpiece. Fashioning the block requires auto body filler or modeling rubber to make a mold of the profile. Start by nailing together a small box slightly larger than the sample molding and at least ¼ inch deeper than the thickest part of the molding. Prepare the filler following the manufacturer's instructions and fill about half the box with it. Lay a single thickness of plastic wrap over the box and, while the filler is still soft, press the molding sample into it *(above, left)* and clamp it firmly in place. Let the filler harden, then remove the molding sample from the box and the nails from the ends. Now saw off both ends of the box. Stretch a piece of sandpaper abrasive-side up across the molded side of the box. Use the molding sample to press the paper against the hardened filler, then staple the ends to the sides of the box *(above, right)*.

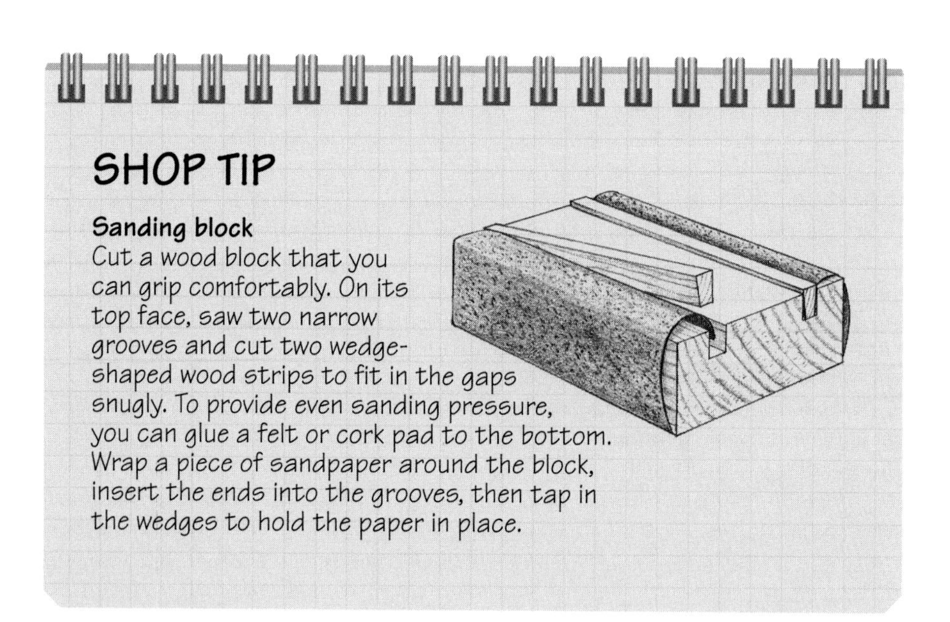

SHOP TIP

Sanding block

Cut a wood block that you can grip comfortably. On its top face, saw two narrow grooves and cut two wedge-shaped wood strips to fit in the gaps snugly. To provide even sanding pressure, you can glue a felt or cork pad to the bottom. Wrap a piece of sandpaper around the block, insert the ends into the grooves, then tap in the wedges to hold the paper in place.

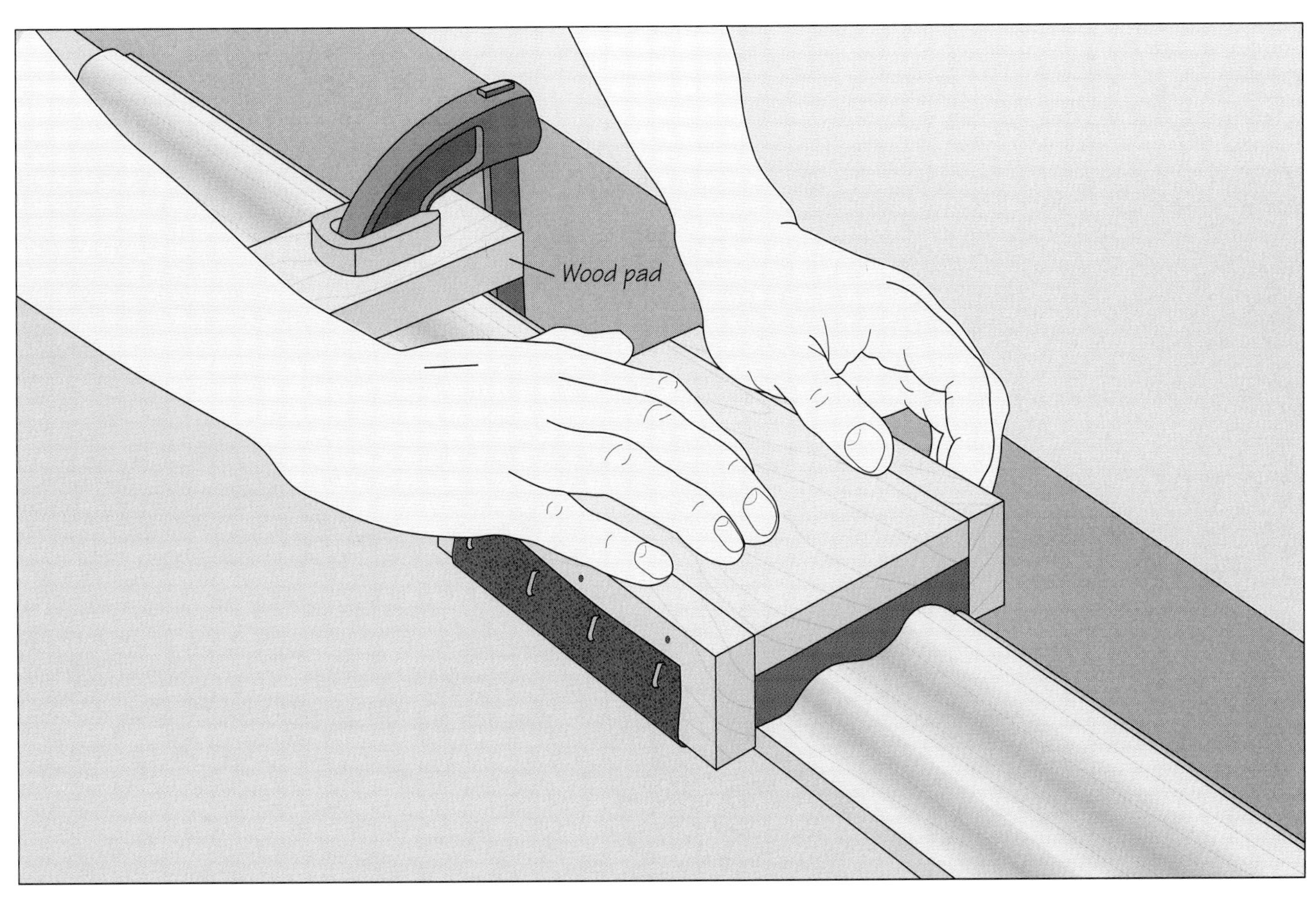

2 Smoothing the molding
Clamp the workpiece to a table, using a wood pad to protect the stock. Slide the block back and forth along the molding *(above)*.

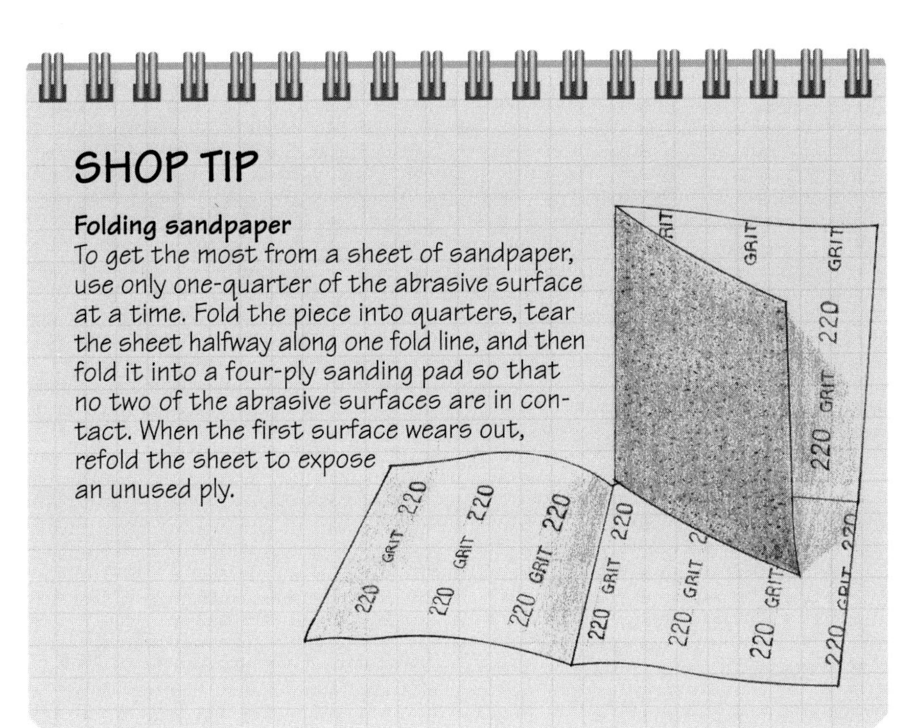

SHOP TIP

Folding sandpaper
To get the most from a sheet of sandpaper, use only one-quarter of the abrasive surface at a time. Fold the piece into quarters, tear the sheet halfway along one fold line, and then fold it into a four-ply sanding pad so that no two of the abrasive surfaces are in contact. When the first surface wears out, refold the sheet to expose an unused ply.

TOOL EXTENSIONS AND TABLES

The jigs in this chapter will allow you to transform portable power tools into small stationary machines and handle large workpieces on your stationary power tools more easily. The stand shown on page 105, for example, converts a plate joiner into a stationary biscuit-slot cutter, a more convenient setup for some operations. The router table on page 106 is essential for some cuts. Featuring an adjustable fence and a storage compartment, it allows you to set up your router as a mini-shaper. In a small shop, a removable router table can be attached to a table saw *(page 107)*. The three-in-one tool table shown on page 108 can convert your router, saber saw, and electric drill into stationary tools.

Extension tables and roller stands greatly expand the versatility of tools like band saws and drill presses. See pages 110 and 111 for jigs that enhance the capacity of these tools to handle large workpieces safely. All the accessories featured in this chapter can be custom-made to suit your needs, at a fraction of the cost of commercial versions.

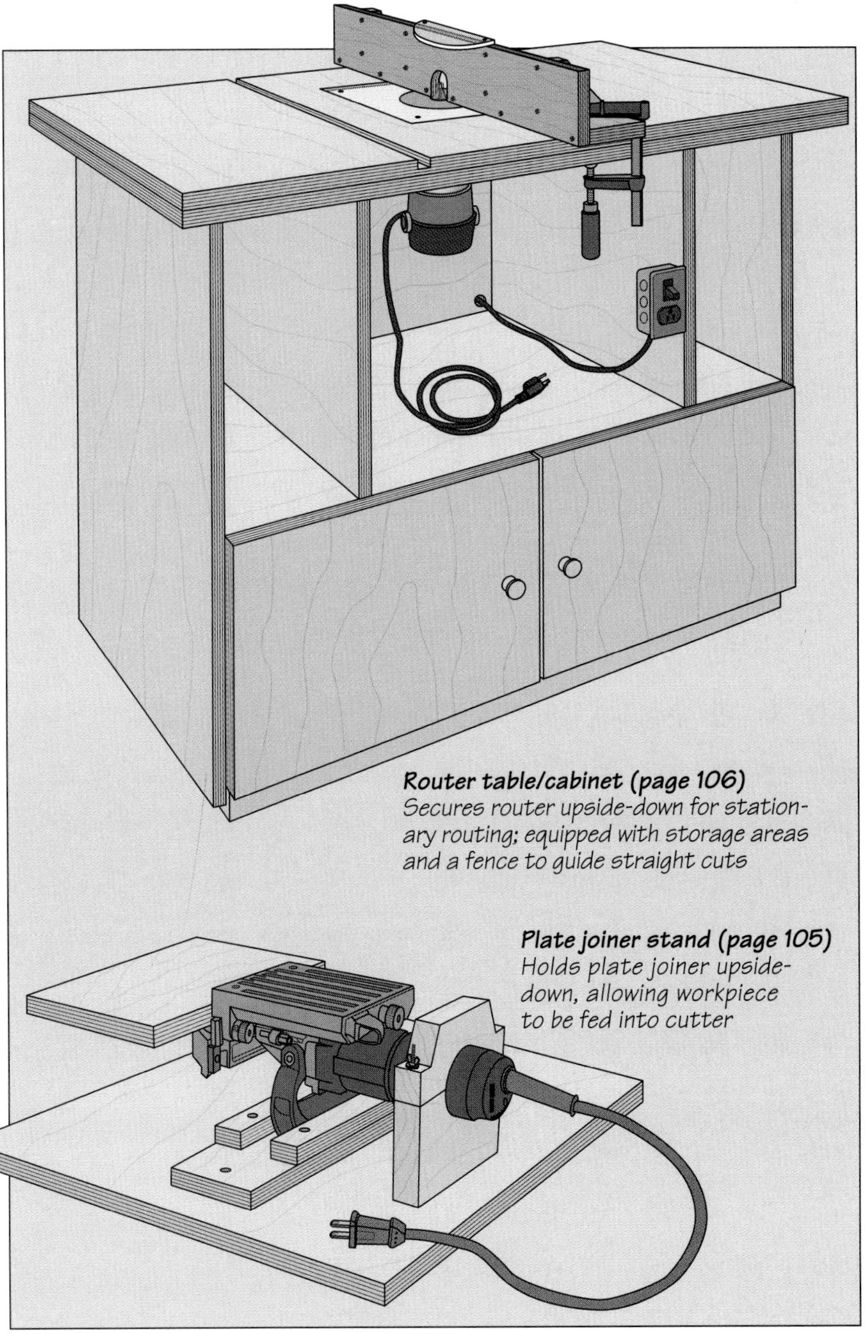

Secured to a band saw table, the plywood extension table at left more than doubles the original table's surface area, allowing better control over a cut—especially with long stock.

Router table/cabinet (page 106)
Secures router upside-down for stationary routing; equipped with storage areas and a fence to guide straight cuts

Plate joiner stand (page 105)
Holds plate joiner upside-down, allowing workpiece to be fed into cutter

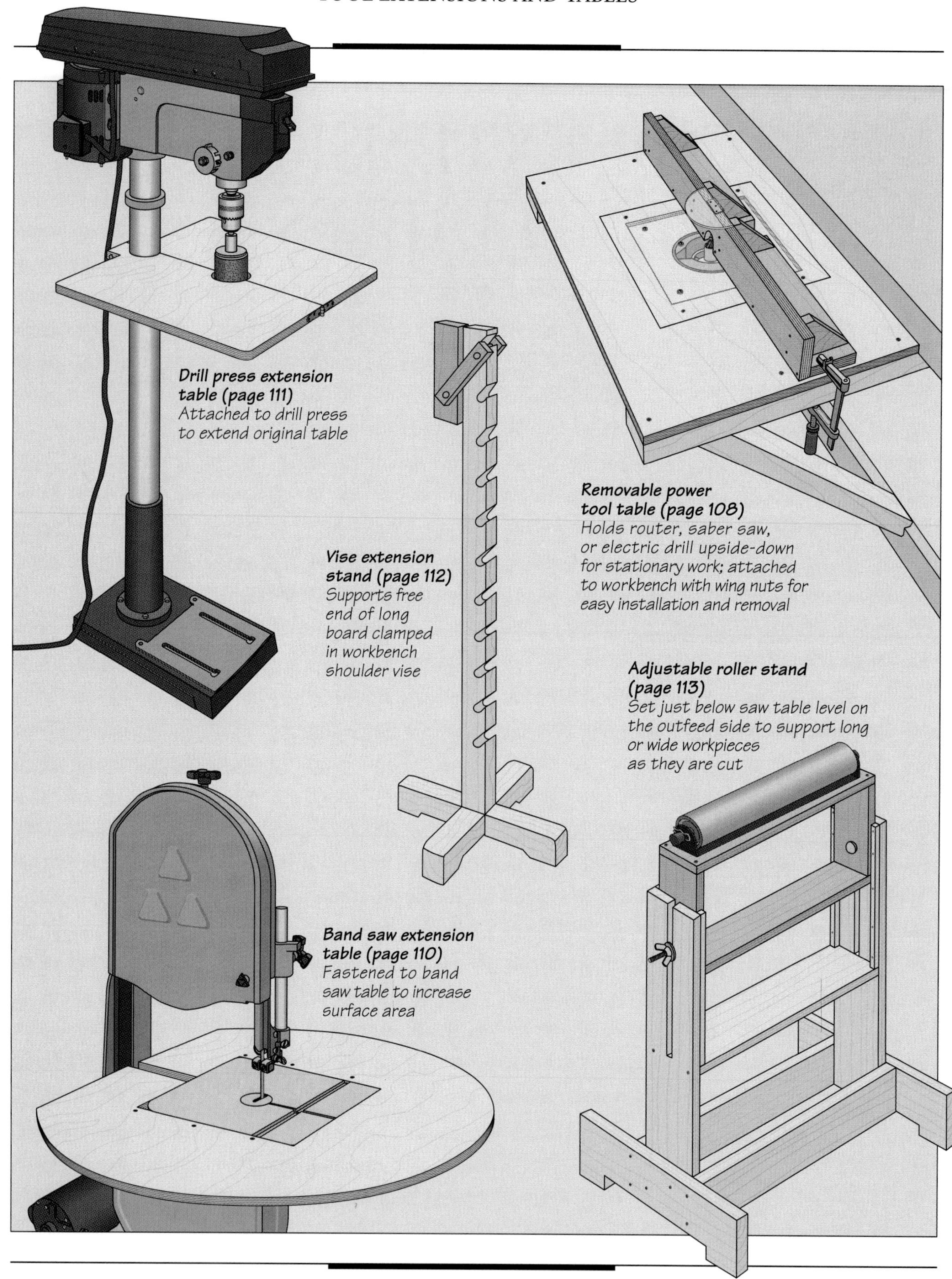
Drill press extension table (page 111)
Attached to drill press to extend original table
Removable power tool table (page 108)
Holds router, saber saw, or electric drill upside-down for stationary work; attached to workbench with wing nuts for easy installation and removal
Vise extension stand (page 112)
Supports free end of long board clamped in workbench shoulder vise
Adjustable roller stand (page 113)
Set just below saw table level on the outfeed side to support long or wide workpieces as they are cut
Band saw extension table (page 110)
Fastened to band saw table to increase surface area

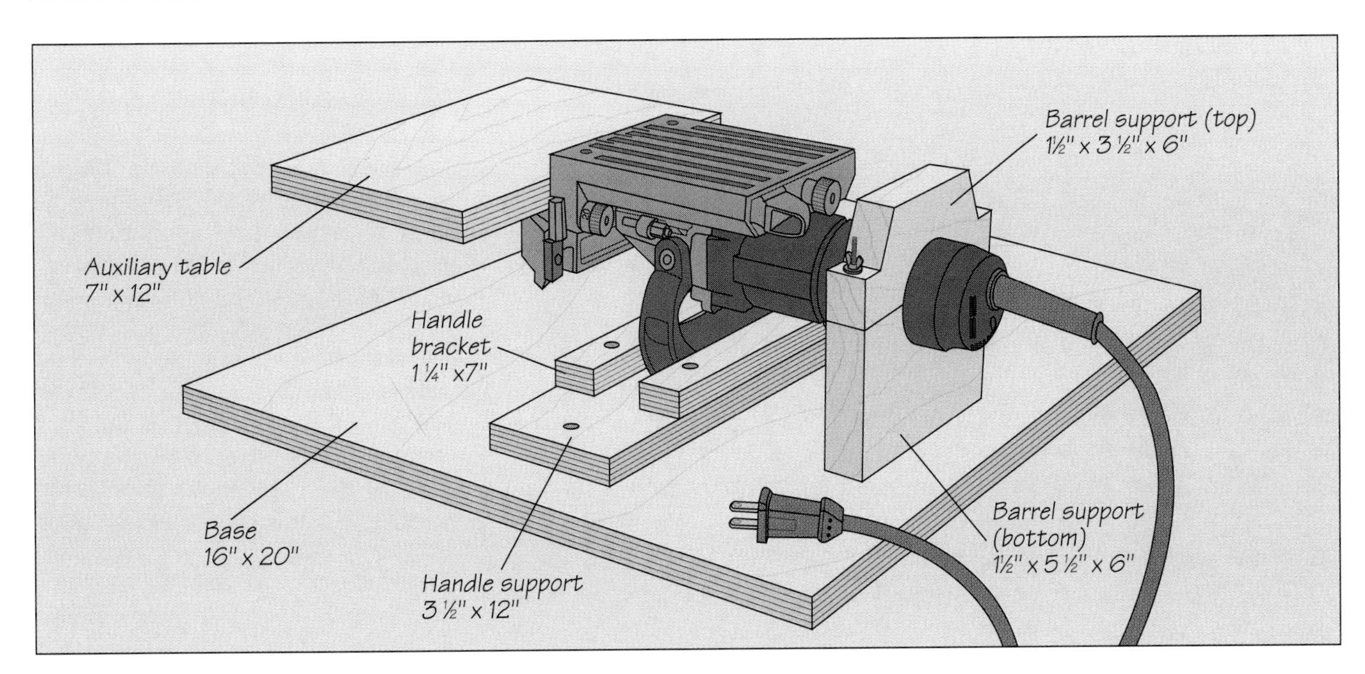

1 Building the jig
Paired with a plate joiner, the jig shown above will reduce the setup time needed to cut slots for biscuits in a series of workpieces. Build the jig from ¾-inch plywood, except for the barrel support, which should be solid wood. Refer to the illustration for suggested dimensions. Screw the handle support to the base and attach the handle brackets, spacing them to fit your tool. With the plate joiner resting upside-down on the handle support, butt the barrel support against the motor housing and outline its shape on the stock. Bore a hole for the barrel and cut the support in two across its width, through the center of the hole. Screw the bottom part to the base and fit the other half on top. Bore holes for hanger bolts through the top on each side of the opening, then drive the bolts into the bottom of the support. Use wing nuts to hold the two halves together. Finally, screw the auxiliary table to the joiner's fixed-angle fence. (It may be necessary to drill holes in the fence for the screws.)

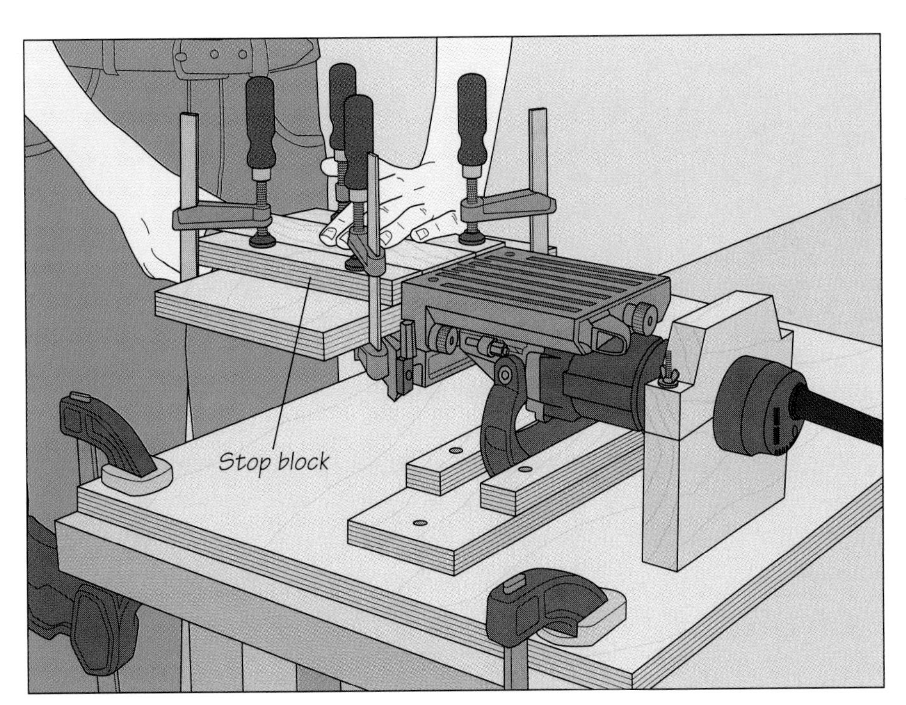

2 Cutting the slots
Secure the plate joiner in the stand and clamp the jig base to a work surface. Set the fence at the correct height and, for repeat cuts, clamp stop blocks to the auxiliary table to center the workpiece in front of the cutter. For each cut, put the workpiece flat on the table and butted against the joiner's faceplate, then turn on the tool and push the stock and the table into the blade *(left)*.

A ROUTER TABLE/CABINET

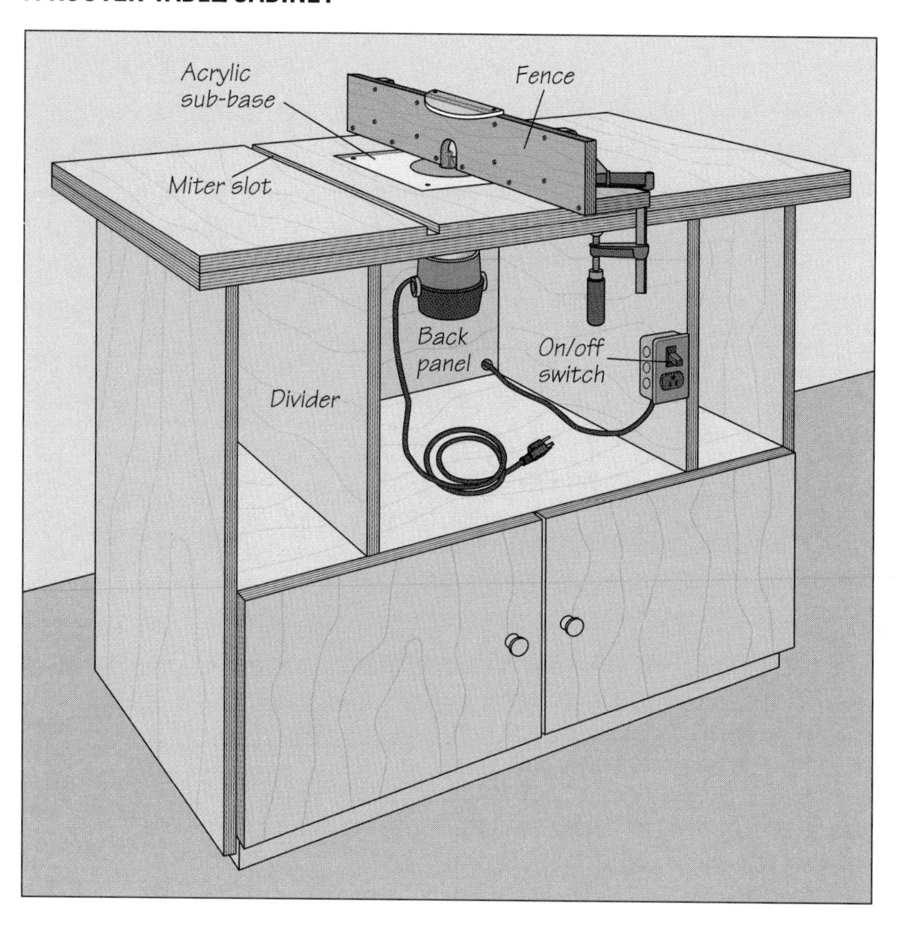

1 Assembling the table
Built entirely from ¾-inch plywood, the table shown at left allows you to use your router as a stationary tool—a requirement of many operations. It features a large top with a slot for a miter gauge, an adjustable fence, a storage shelf, and cupboards. Start with the basic structure of the table, sizing the bottom, sides, back, shelf, dividers, and doors to suit your needs. Fix these parts together, using the joinery method of your choice. The table shown is assembled with biscuit joints and screws. Bore a hole through the back panel to accommodate the switch's power cord. For the top, cut two pieces of plywood and use glue and screws to fasten them together; the pieces should be large enough to overhang the sides of the cabinet by 2 or 3 inches. Fix the top to the cabinet. Finally, fasten a combination switch-receptacle to one of the dividers, with a power cord long enough to reach a nearby outlet. When you use the table, plug in the router and leave its motor on. Use the table's switch to turn the tool on and off.

2 Preparing the tabletop
The router is attached to the top with a square sub-base of ¼-inch-thick clear acrylic. Several steps are necessary to fit the sub-base to the top and then to the router. First, position the sub-base at the center of the top and outline its edges with a pencil. Mark the center of the sub-base and drill a pilot hole through the acrylic and the top. Remove the sub-base and rout out a ¼-inch-deep recess within the outline *(right)*. Use a chisel to pare to the line and square the corners. Then, using the pilot hole as a center, cut a hole through the top to accommodate your router's base plate. Next, use a straightedge guide to help you rout the miter slot across the top: Clamp the guide square to the front edge of the top and butt the router against it as you plow a slot that is just wide enough to fit your miter gauge bar snugly.

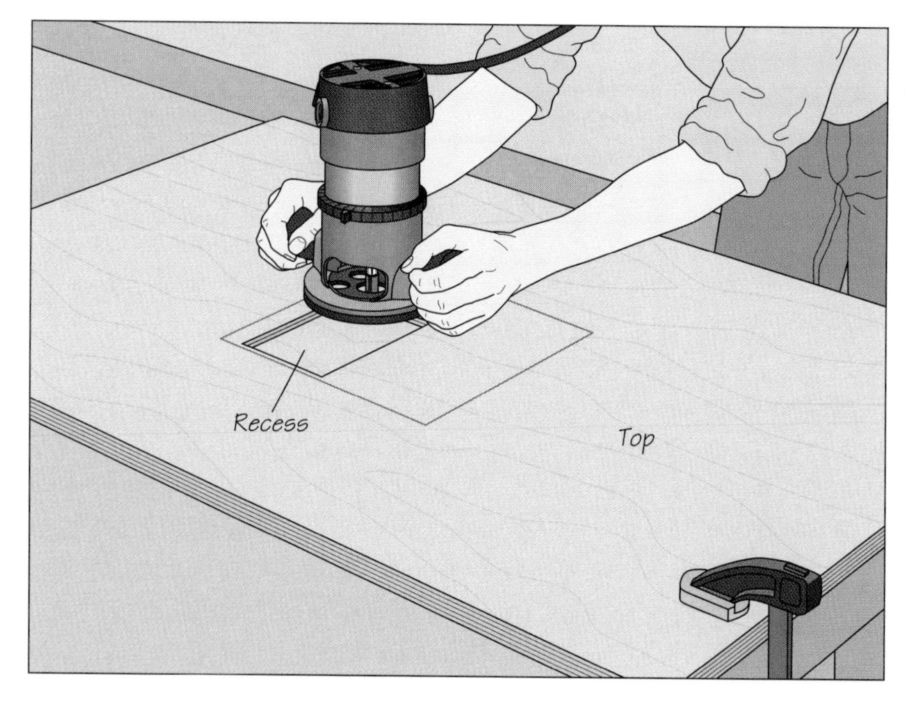

3 Preparing the sub-base
Drill a hole through the center of the sub-base slightly larger than your largest router bit, and fasten the sub-base to the router using flat-head machine screws *(above)*. Set the sub-base in the table recess and attach it with wood screws, drilling pilot holes first and countersinking all fasteners.

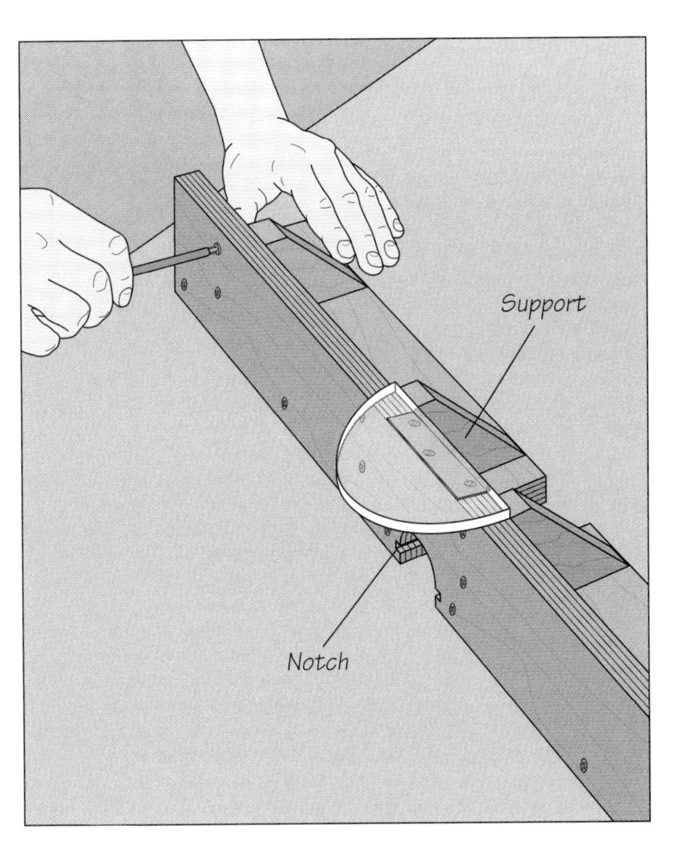

4 Making the fence
Cut two pieces of 3/4-inch plywood and screw them together in an L shape. Saw a notch out of the fence's bottom edge to accommodate your largest bit, and screw four triangular supports to the back for added stability *(above)*. Attach a clear semicircular plastic guard, with a hinge so it can be swung out of the way. To use the router table for a straight cut, clamp the fence in position and feed the workpiece into the bit, holding it flush against the fence.

SHOP TIP

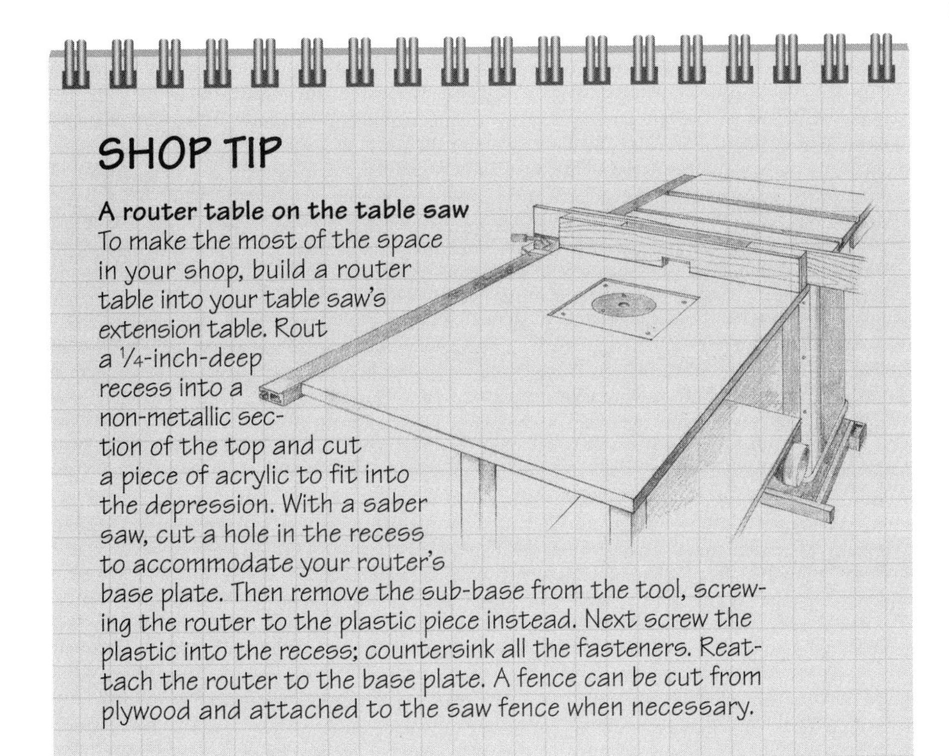

A router table on the table saw
To make the most of the space in your shop, build a router table into your table saw's extension table. Rout a 1/4-inch-deep recess into a non-metallic section of the top and cut a piece of acrylic to fit into the depression. With a saber saw, cut a hole in the recess to accommodate your router's base plate. Then remove the sub-base from the tool, screwing the router to the plastic piece instead. Next screw the plastic into the recess; countersink all the fasteners. Reattach the router to the base plate. A fence can be cut from plywood and attached to the saw fence when necessary.

A REMOVABLE POWER TOOL TABLE

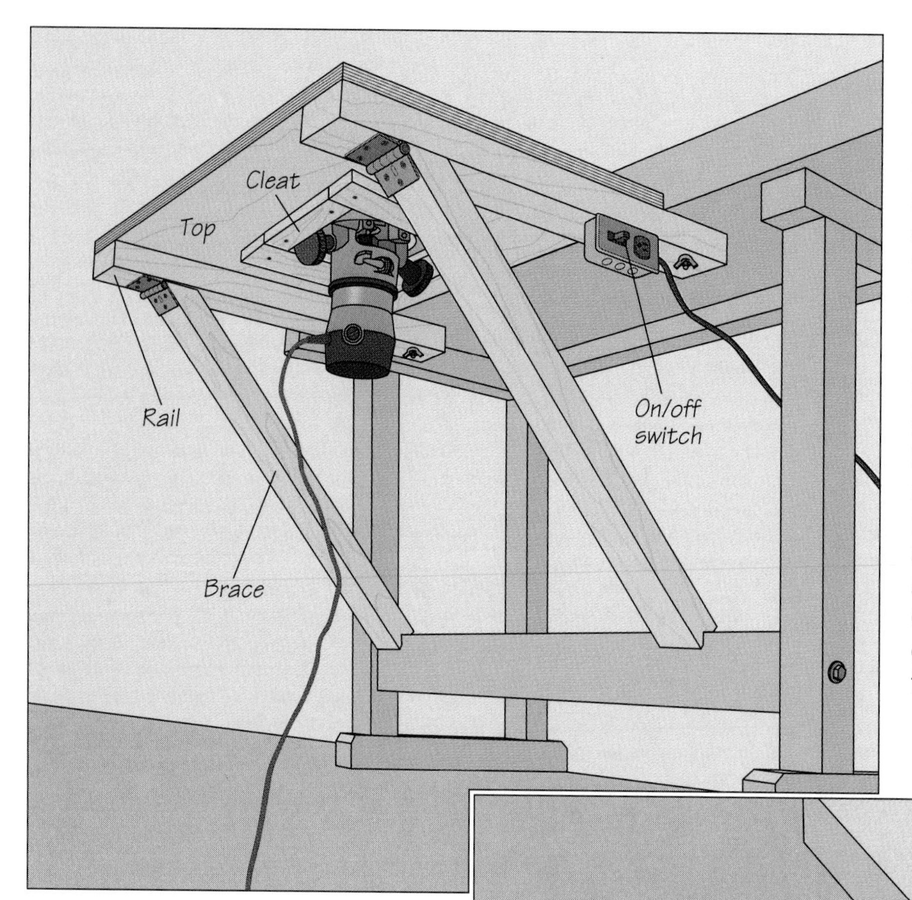

1 Building the table
Attached to a workbench or table, the removable extension table shown at left allows you to convert three different portable power tools into stationary devices: the router, the electric drill, and the saber saw. Size the parts according to your needs. Start by cutting the top from ¾-inch plywood, and the rails and braces from 2-by-4 stock. Saw the rails a few inches longer than the width of the top so they can be fastened to the underside of the bench using wing nuts and hanger bolts. The hinged braces should be long enough to reach from the underside of the rails to a leg stretcher on the bench. Cut a bevel at the top end of the braces and a right-angled notch at the bottom. Mount an On/off switch for the tool on the underside of one of the rails as you would for a table/cabinet *(page 106)*.

Cleat

2 Preparing the tabletop
Cut a rectangular hole out of the tabletop's center the same size as the inserts you will use for the tools *(page 109)*. Then screw cleats to the underside of the top, forming a ledge to which the inserts can be fastened *(right)*.

3 Preparing the router insert

Saw the three tool inserts, sizing them precisely to fit in the tabletop hole. An acrylic router sub-base is fitted into the insert following the directions for the router table on pages 106 and 107. To install the insert in the table, set it on the cleats and screw the insert to the cleats at each corner *(right)*. Drill pilot holes for the screws. Make a fence as you would for the router table on page 107. Secure the fence in the desired position with clamps.

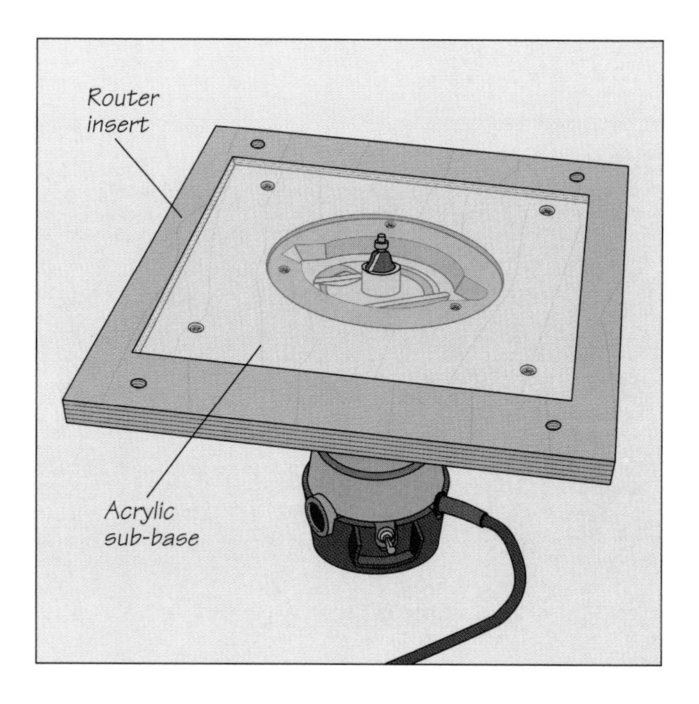

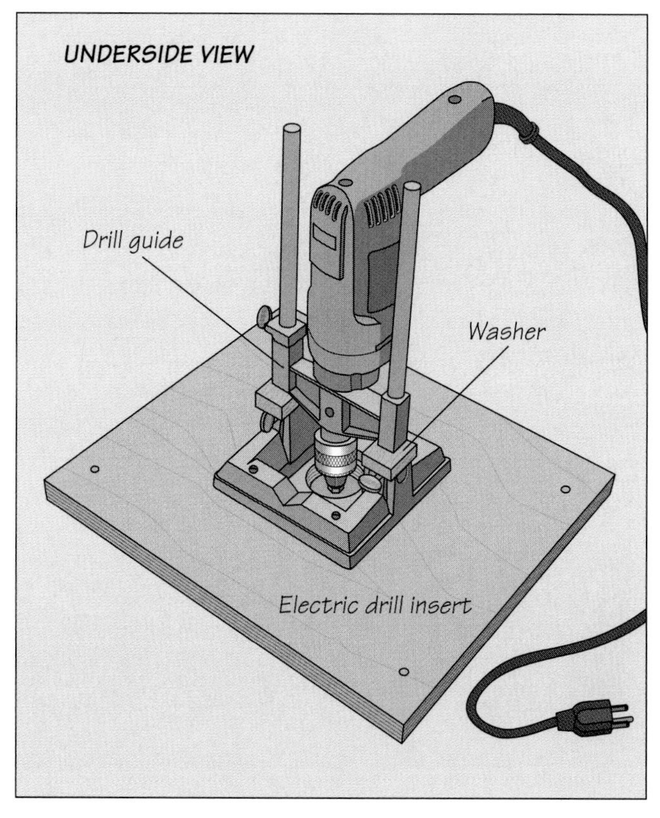

4 Preparing the electric drill insert

Bore a hole through the center of the drill insert slightly wider than the largest sanding drum or other accessory you plan to use. (The stationary drill is particularly useful for sanding.) Screw a commercial drill guide to the underside of the insert with the drill chuck centered over the hole *(above)*. (You may need to drill holes in the base of the drill guide to fasten it in place.) The bit or accessory in the drill chuck should protrude from the top of the insert without the chuck being visible. Place wooden washers under the guide rods of the drill guide to adjust the height of the drill, if necessary. Fasten the insert to the cleats as you did the router insert.

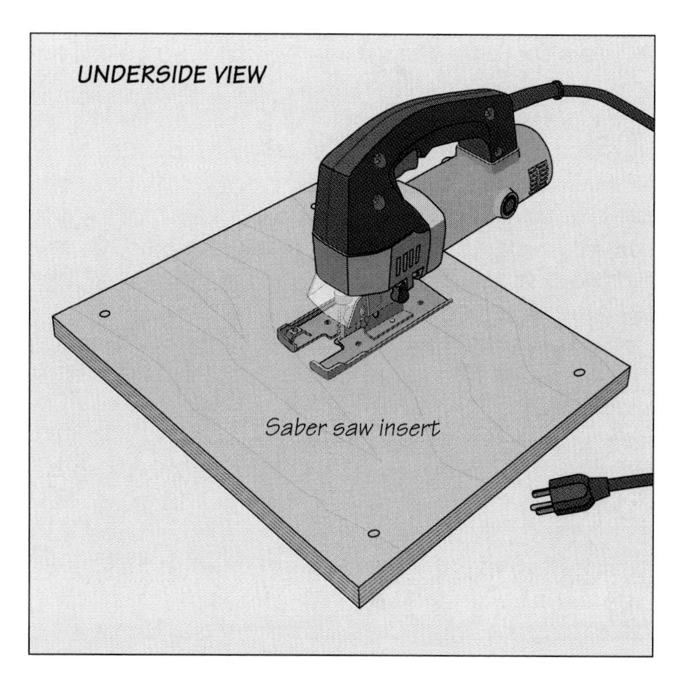

5 Preparing the saber saw insert

Position the saw's base plate so the blade will be in the center of the insert and mark its location. Bore a hole at the mark large enough to clear the blade. Screw the saw's base plate to the insert *(above)*. If there are fewer than four screw holes in the base plate, drill additional holes. Mount the insert to the cleats.

A BAND SAW EXTENSION TABLE

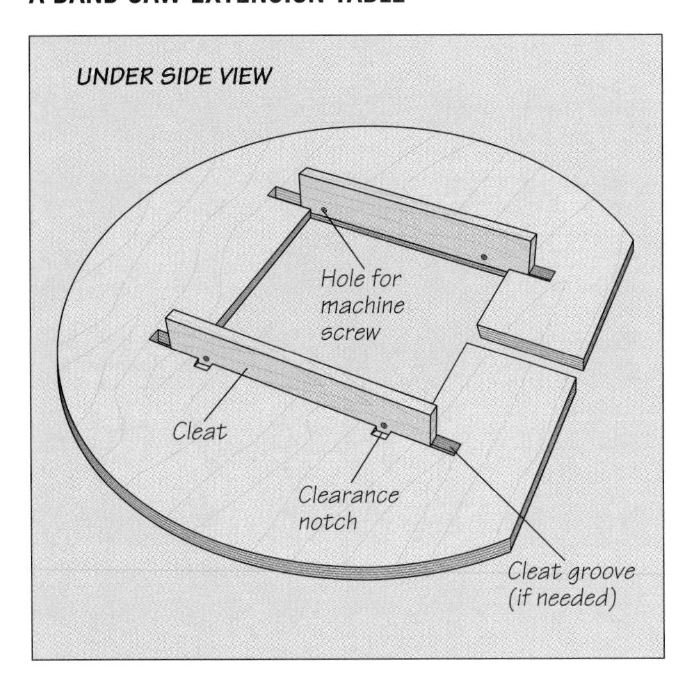

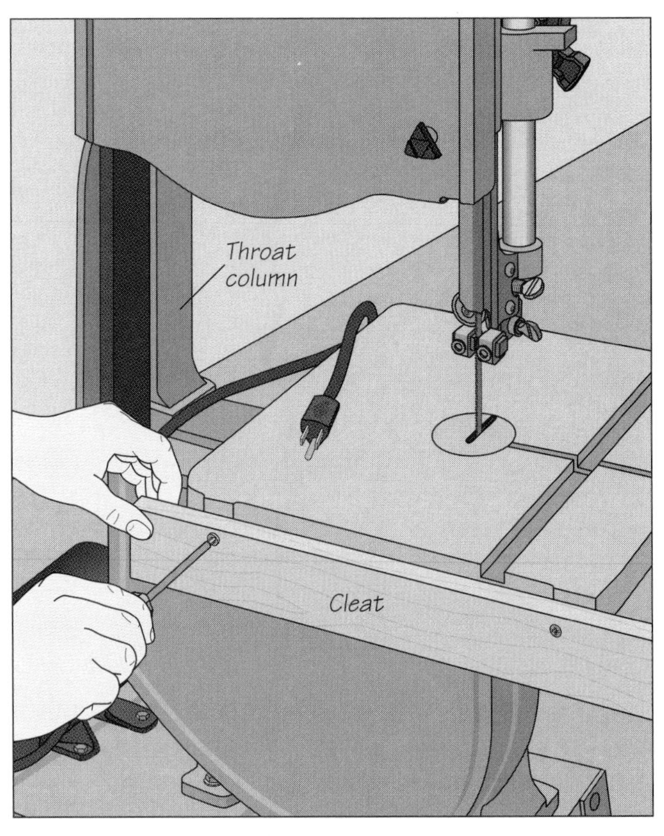

Building and installing a table for the band saw

An extension table on your band saw will enable you to cut long or wide pieces with greater ease and control. Using ¾-inch plywood, cut the jig top to a suitable diameter, then saw out the center to fit around the saw table and trim a portion of the back edge to clear the throat column. Cut a 1½-inch-wide channel from the back of the table to the cutout so the table can be installed without removing the blade. Next, prepare two cleats that will be used to attach the jig to the saw table. For these, two 1-by-3s should be cut a few inches longer than the saw table. Position them against the sides of the saw table so that they are ¾ inch below the table surface, with at least ¼ inch of stock above the threaded holes. (Make sure your machine has these holes; most band saws have them for mounting an accessory rip fence.) Depending on the position of the holes on your saw table, you may have to position the top of the cleats closer than ¾ inch to the top of the saw table. In that case you will have to rout grooves for the cleats on the underside of the top to allow the tabletop to sit flush with the machine's table *(above)*. Mark the hole locations on the cleats, bore a hole at each spot, and fasten the cleats to the table with the machine screws provided for the rip fence *(right, top)*. Then place the tabletop on the cleats and screw it in place *(right, bottom)*; be sure to countersink the screws. The top should sit level with the saw table. To remove the jig, remove only the machine screws, leaving the cleats attached permanently to the top. You may need to cut clearance notches in the underside of the top so that you can reach the machine screws.

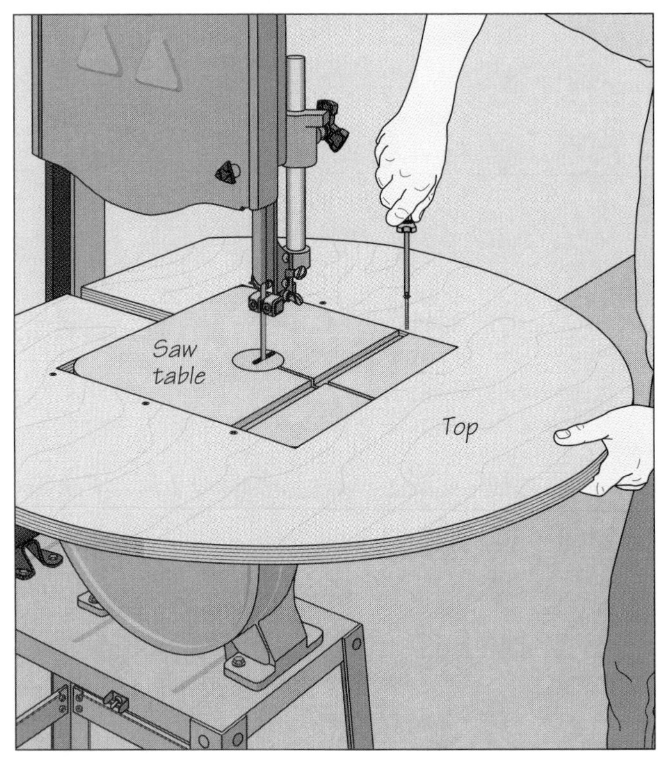

A DRILL PRESS EXTENSION TABLE

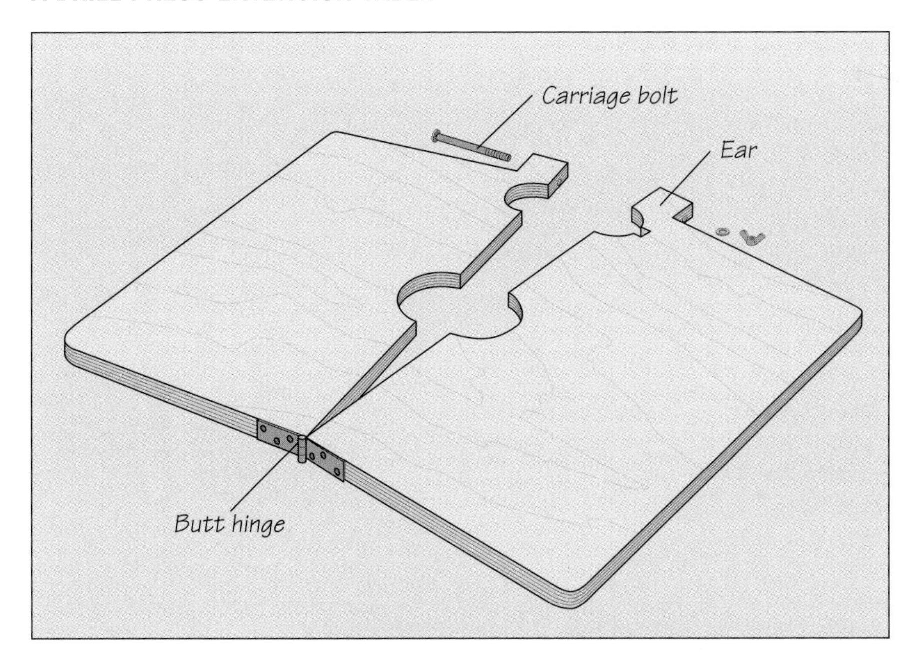

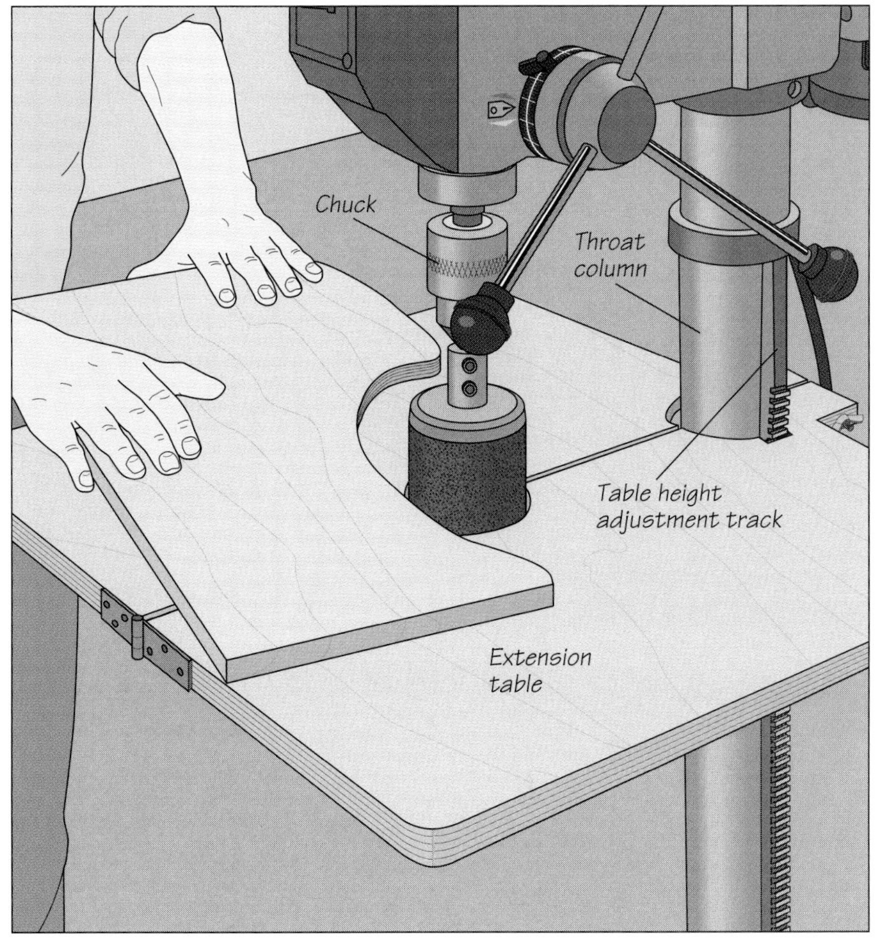

Fitting a drill press with an extension table
The small table typical of most drill presses will not adequately support large workpieces, especially when the tool is set up for sanding operations. To build an extension table, start by cutting a piece of ¾-inch plywood into a square with dimensions that suit your needs. Then mark a line down the middle of the piece and draw two circles centered on the line. Locate the first about 4 inches from the back edge, sizing it to fit snugly around the drill press column. Locate the second hole under the chuck; make its diameter about ½ inch greater than the largest accessory you plan to use. To help pinpoint the center of the hole, install a bit in the chuck and measure the distance from the column to the bit. Prepare to install the jig on the drill press table by cutting its back edge, leaving a rectangular "ear" that protrudes behind the back hole. Bore a hole through the ear for a ¼-inch-diameter carriage bolt. Next, saw the jig in half along the centerline and cut out the two circles. You may need to make other cuts to clear protrusions on your machine. On the model shown, a notch was needed for the table height adjustment track on the throat column. Finally, screw a butt hinge to the front edge of the jig to join the two halves as shown. The carriage bolt and wing nut will clamp the table in place on top of the drill press table.

VISE EXTENSION STAND

Making and using the stand

The vise extension stand shown at right is used to support the free end of a long board clamped in the shoulder vise of a workbench. The dimensions given in the illustration will work well with most benches. Cut the upright to length, then saw angled notches at 2½-inch intervals along its length, starting 5 inches from the bottom. Make the notches about 1 inch long and ½ inch wide. Then saw the feet to length and cut recesses along their bottom edges. Join the feet with an edge half-lap joint: Cut half-laps in the top edge of one foot and the bottom edge of the other, then glue the two together. Once the adhesive is dry, use a lag screw to attach the upright to the feet; drive the screw into the upright from underneath the feet. Cut the support piece and swivel bars, angling the top of the support piece about 10° toward the upright. To join the support piece to the swivel bars, bore holes for ⅜-inch-diameter dowels through the square end of the support piece and near the ends of the bars, and slip the dowels into the holes; glue the dowels to the bars, but leave the support piece free to pivot. To use the stand, insert the dowel at the top end of the swivel bars in the appropriate slot in the upright for the height you need and prop your workpiece on the support piece.

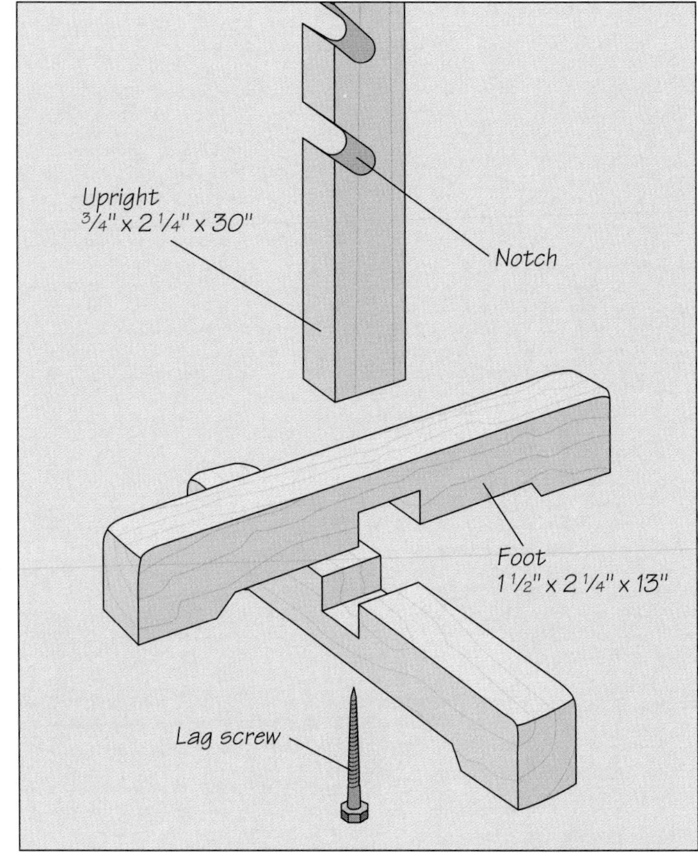

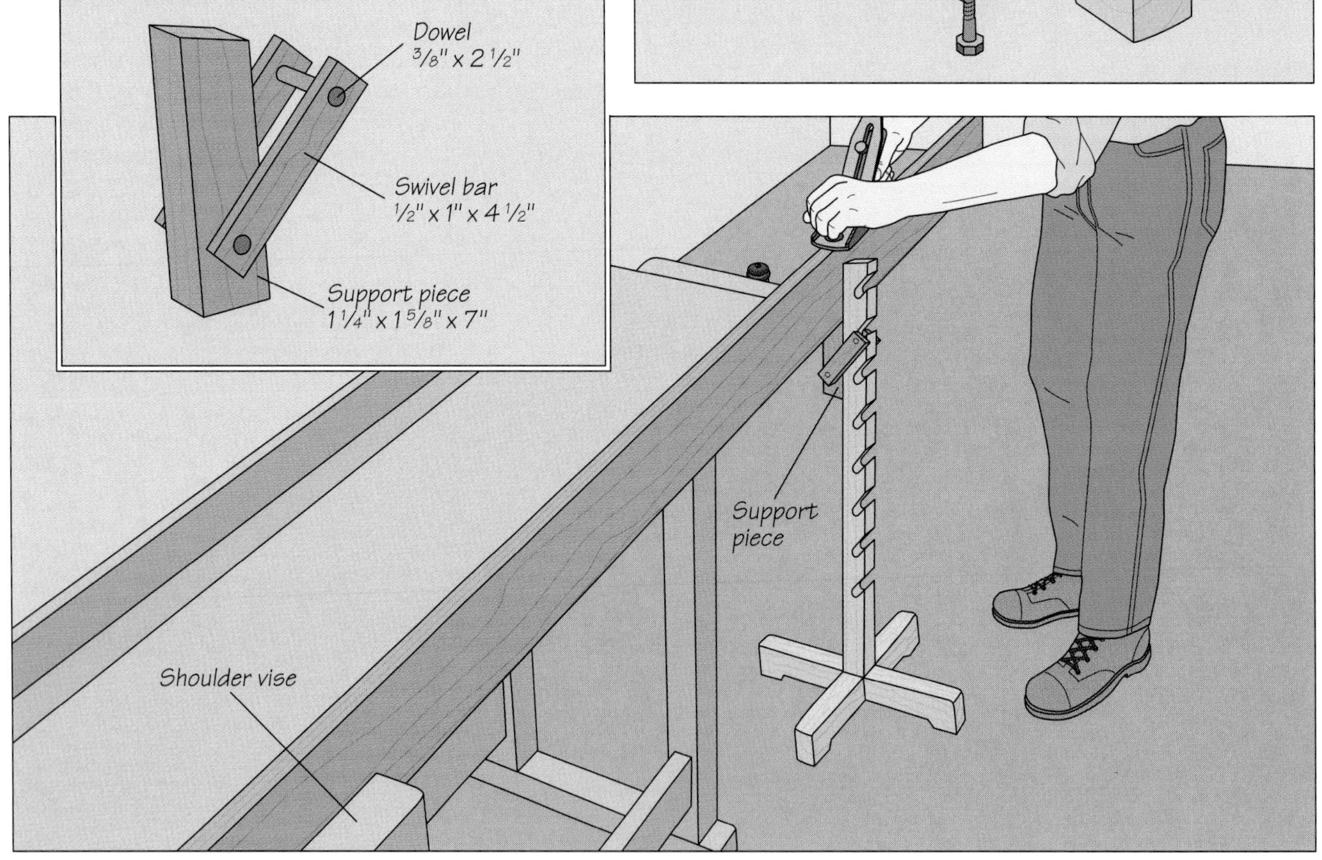

ADJUSTABLE ROLLER STAND

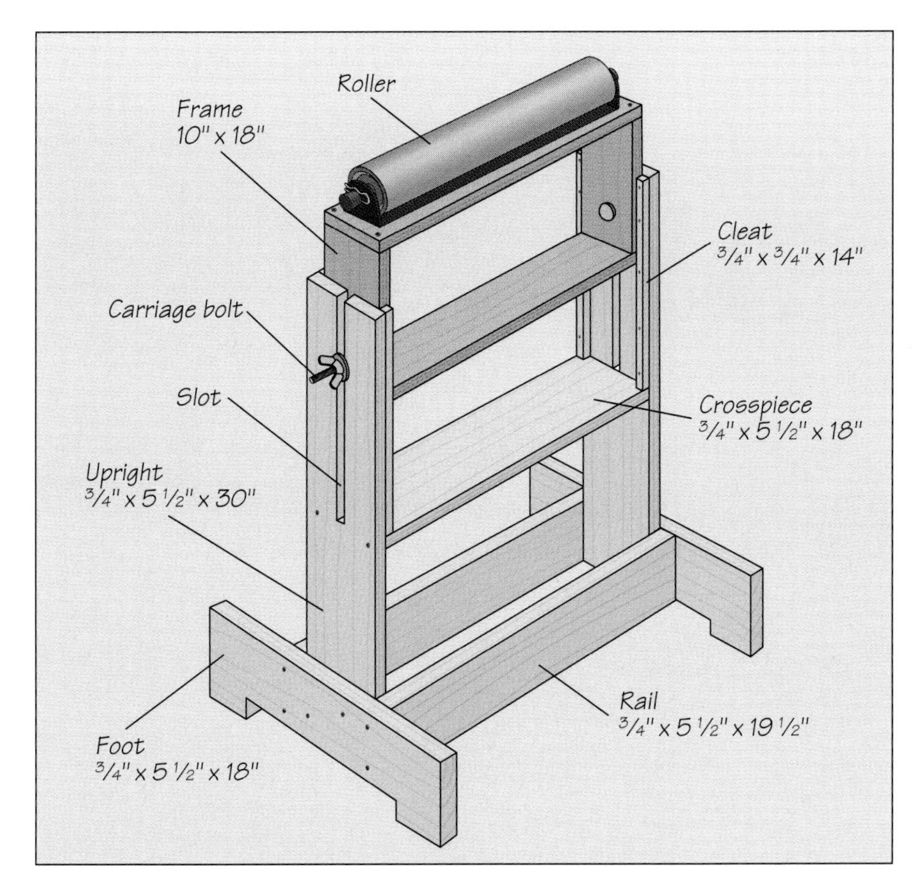

Building the stand

Set up on the outfeed side of a saw table, an adjustable roller stand like the one shown at left will help you support large workpieces. Start by constructing the frame for the roller, cutting the four pieces from 1-by-4 stock. Glue the frame together with butt joints, adding screws to reinforce the connections. Then bore a hole in each side for a 1/4-inch-diameter carriage bolt. Center the hole 3 inches from the bottom of the frame. Screw the roller to the top. Cut the remaining pieces of the stand from 1-by-6 stock, referring to the dimensions provided, and rout a 14-inch-long, 1/4-inch-wide slot down the middle of the two uprights. Screw the crosspiece to the uprights, aligning the top of the piece with the bottom of the slot. Fasten the upright and rails to the feet. To guide the roller frame, nail 1-by-1 cleats to the uprights about 1/4 inch in from the edges. To set up the stand, position the frame between the uprights, fitting the carriage bolts into the slots from inside the frame. Slip washers over the bolts and tighten the wing nuts to set the height of the roller slightly below the level of your saw table.

SHOP TIP

A temporary roller stand

Using only a sawhorse, two C clamps, some wood, and a commercial roller, you can make a simple but effective stand to support the outfeed from any of your machines. Make a T-shaped mast for the roller, ensuring it is long enough to hold the roller at a suitable height. Screw the roller to the horizontal member of the mast. Add a brace to one end of the horse for clamping the mast in place: Cut 2 1-by-4s to span the legs and screw them to the legs as shown. To secure the roller stand to the sawhorse, clamp the mast to the braces, making sure the roller is horizontal.

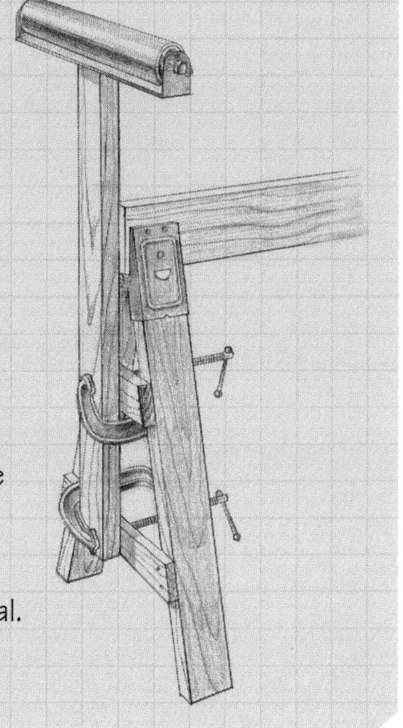

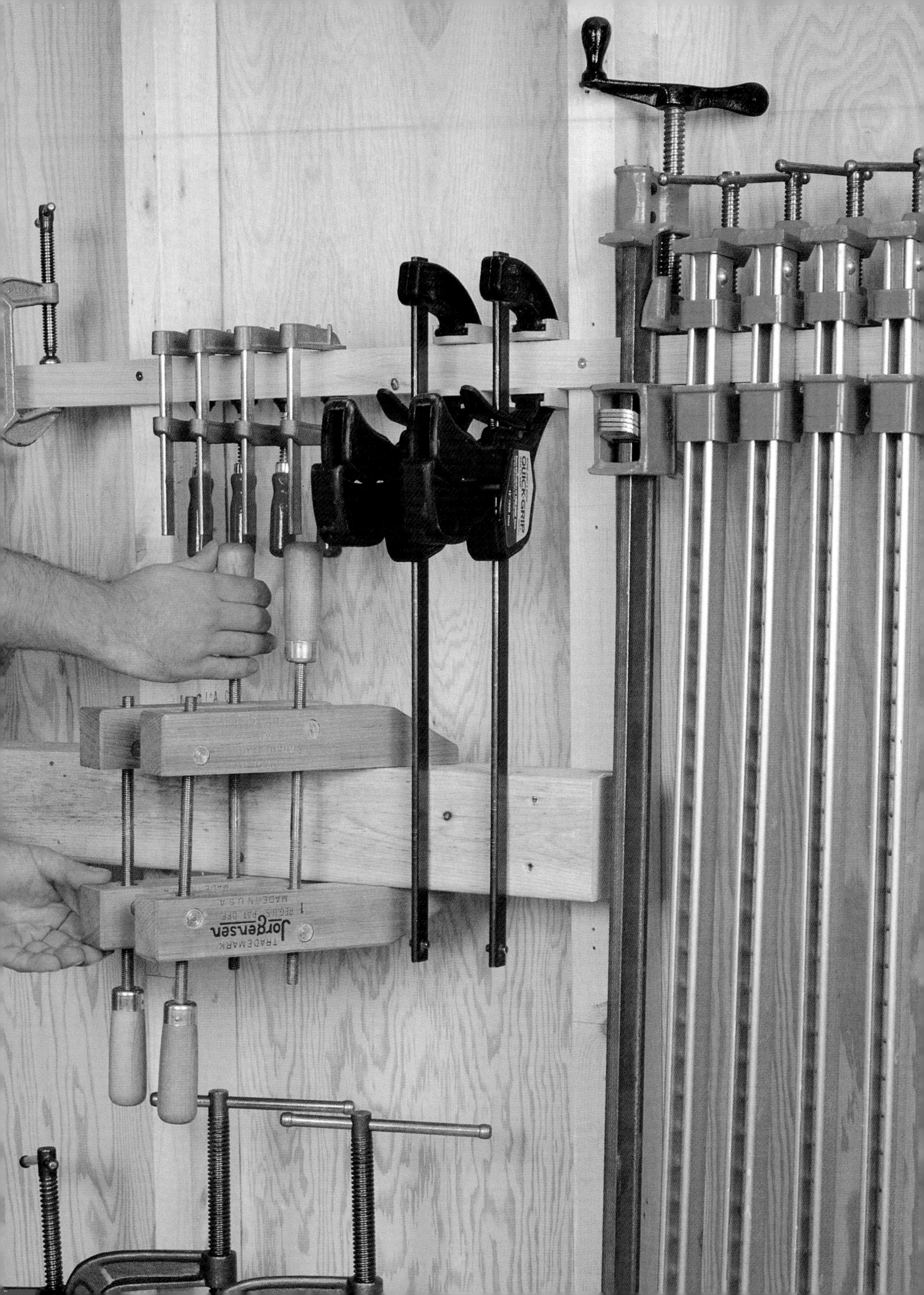
QUICK-GRIP
TRADEMARK
Jorgensen

STORAGE DEVICES

Whether your workshop is tucked away in a corner of your basement or spread out over a two-car garage, storing the tools and materials that accumulate is a persistent challenge. This chapter offers several simple storage devices that can help you win the ongoing battle against clutter. They will keep your tools and materials within easy reach when they are needed, and out of the way when they are not.

For storing hand tools, consider the handsaw holder *(page 117)* and the tool tray, the chisel, and router bit racks shown opposite. As well as helping to organize your tools, these devices will prevent damage to cutting edges.

An effective system for storing clamps is a must. The rack shown on page 118 will accommodate a large selection and, because it is mounted on casters, the rack can be moved to wherever it is needed. Lumber and plywood storage racks are shown on pages 120 and 121. Each of the examples shown can hold enough stock for several projects while conserving precious floor space.

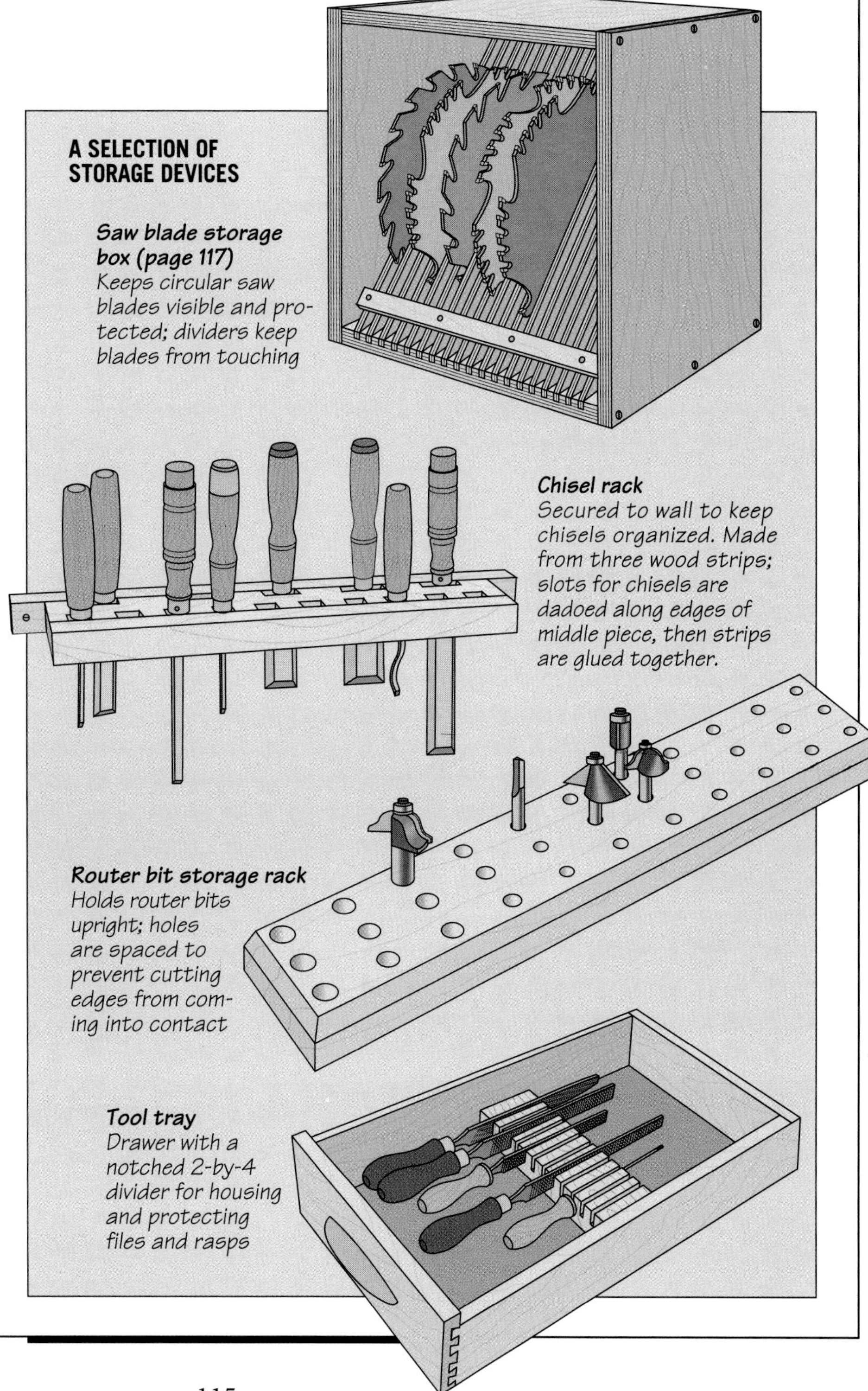

Clamps can be a nuisance to store because of their sheer number in the shop. The simple rack shown at left is made of strips of wood mounted on wall studs. The lower strip is thicker, to keep the clamps leaning toward the wall.

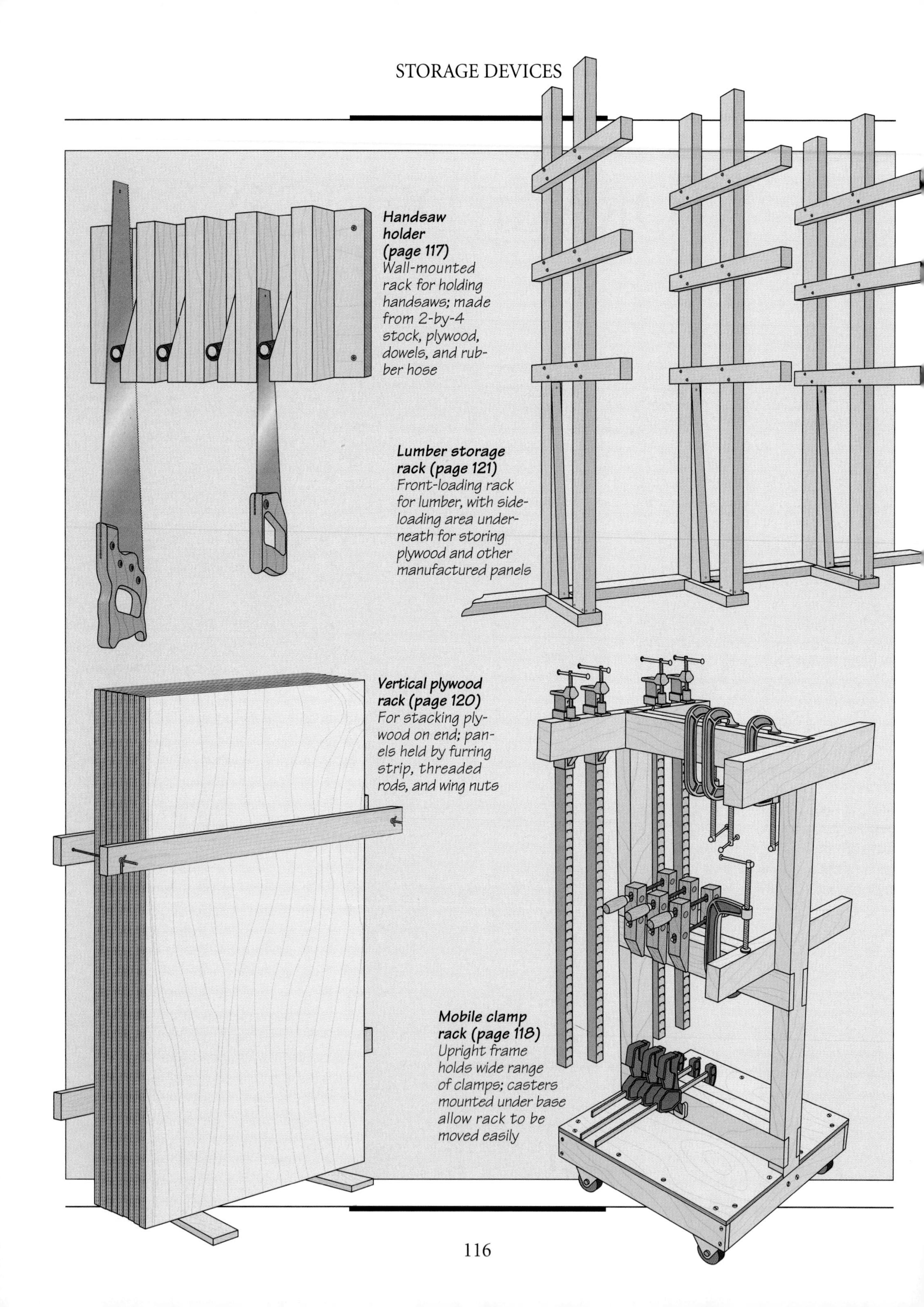
Handsaw holder (page 117)
Wall-mounted rack for holding handsaws; made from 2-by-4 stock, plywood, dowels, and rubber hose
Lumber storage rack (page 121)
Front-loading rack for lumber, with side-loading area underneath for storing plywood and other manufactured panels
Vertical plywood rack (page 120)
For stacking plywood on end; panels held by furring strip, threaded rods, and wing nuts
Mobile clamp rack (page 118)
Upright frame holds wide range of clamps; casters mounted under base allow rack to be moved easily

A SAW BLADE STORAGE BOX

Storing circular saw blades
Organize your circular saw blades in a custom-made storage box like the one shown at right. Build the box from ¾-inch plywood, cutting it a few inches larger than your largest blade and wide enough to hold all your blades. Make the bottom and back from the same piece, routing the dadoes for the dividers first and then cutting the piece in two. This will ensure perfect alignment of the dadoes. Cut the dadoes ¼ inch deep and wide, spaced at ½-inch intervals. Make the dividers out of ¼-inch plywood or hardboard. Fit the dividers in their dadoes, glue them to the bottom and back, then screw the box together. To keep the blades from rolling out of the box, cut a batten from scrap stock and nail it to the dividers near the bottom of the box.

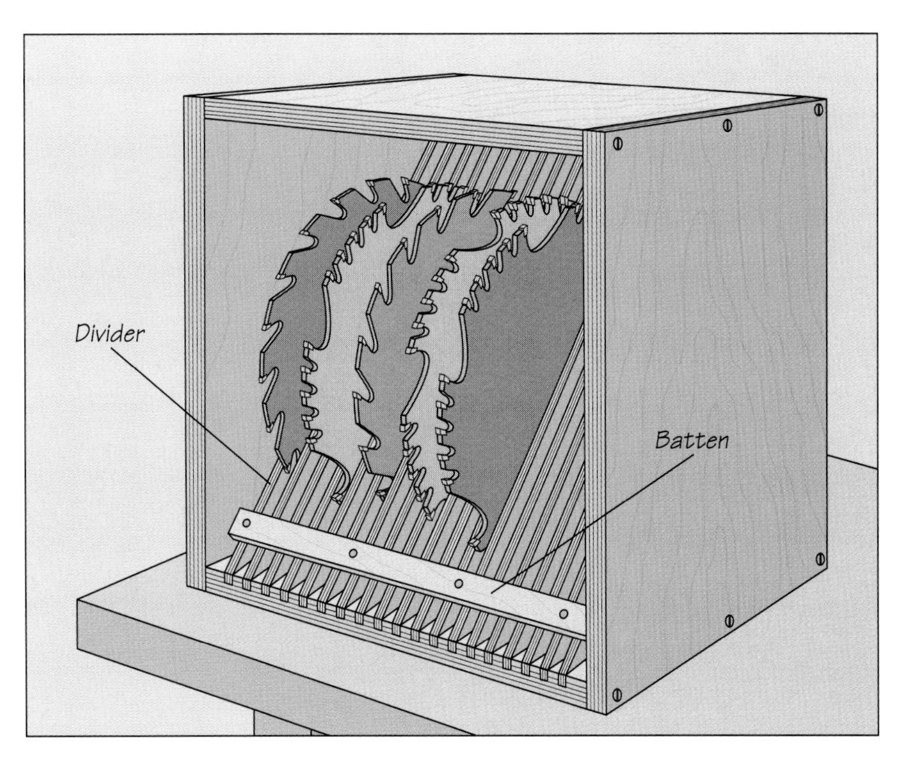

A HANDSAW HOLDER

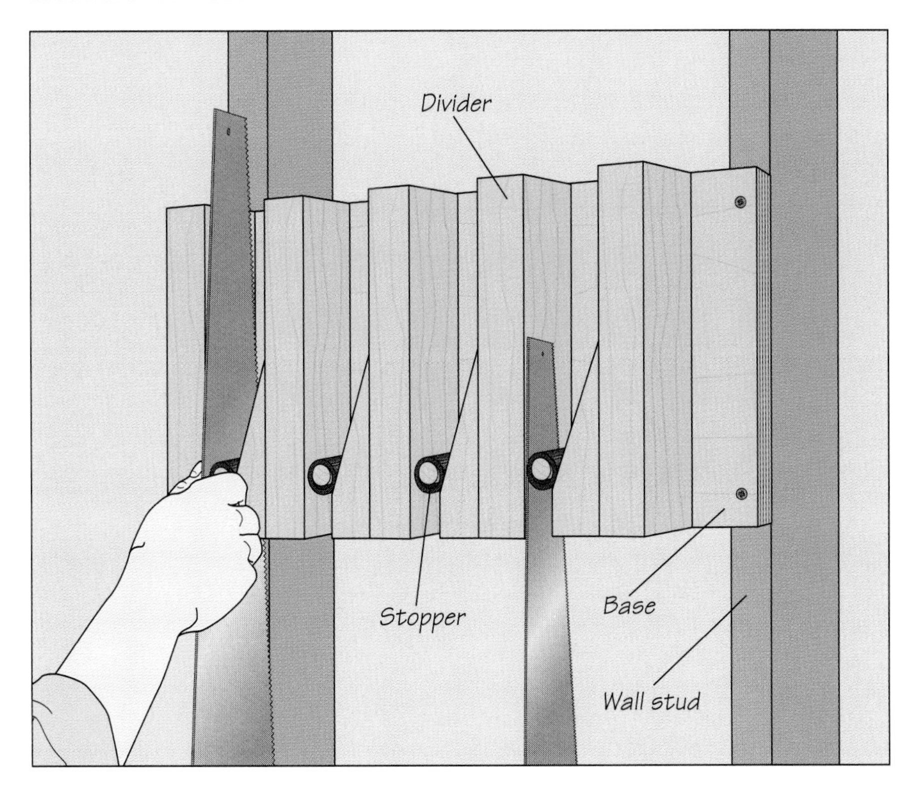

Building and mounting the holder
The wall-mounted rack shown at left, made with a few wood scraps, doweling, and some rubber hose, uses friction to hold your handsaws in place. Cut the base from ¾-inch plywood. Each divider is made from a 10-inch-long 2-by-4 face-glued to a short, tapered 2-by-4. Begin by screwing a 2-by-4 at the left end of the base, then secure the dividers in place, leaving a ½-inch gap between them. The stoppers are 4-inch lengths of ½-inch dowel pressed into slightly shorter pieces of rubber garden hose; use hose with ridges rather than smooth hose. Slip a saw into the rack from below, then tug down on the handle. The stopper will pinch the blade in place. Mark the dowel's position, bore a ½-inch-diameter stopped hole into the base at the mark, then screw the dowel in place. Once all the stoppers are in position, mount the holder to the wall.

MOBILE CLAMP RACK

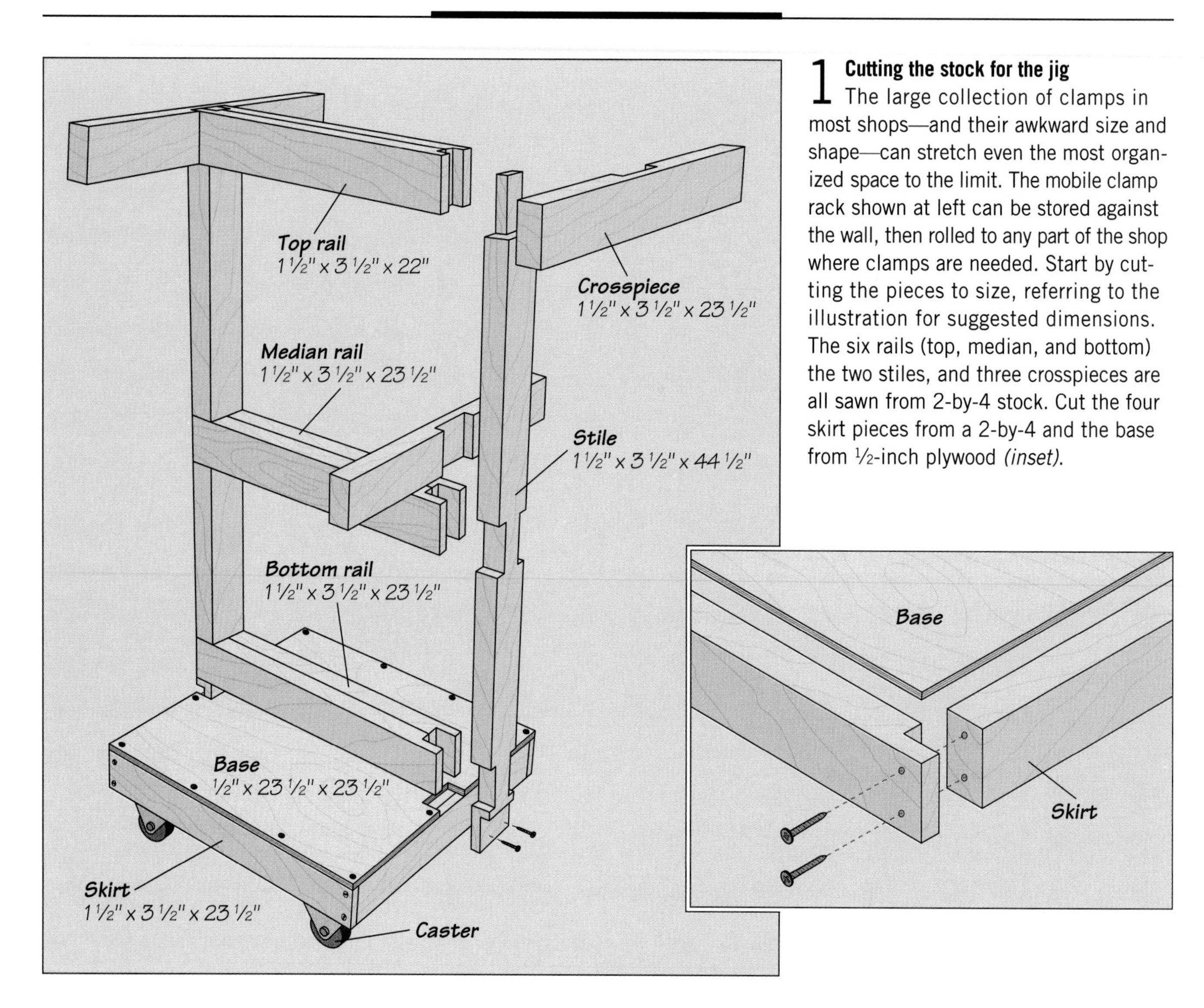

1 Cutting the stock for the jig
The large collection of clamps in most shops—and their awkward size and shape—can stretch even the most organized space to the limit. The mobile clamp rack shown at left can be stored against the wall, then rolled to any part of the shop where clamps are needed. Start by cutting the pieces to size, referring to the illustration for suggested dimensions. The six rails (top, median, and bottom) the two stiles, and three crosspieces are all sawn from 2-by-4 stock. Cut the four skirt pieces from a 2-by-4 and the base from ½-inch plywood *(inset).*

2 Attaching the rails to the stiles
Prepare the rails for the joinery by cutting end rabbets that will fit into notches and dadoes in the stiles. The rabbets should be 1½ inches wide and ¾ inch deep, except for the top rails, which require a rabbet only 1 inch wide. Notch the top end of each stile on three sides, then rout back-to-back dadoes near the bottom end and middle of the stiles; make the dadoes 3½ inches wide and ¾ inch deep. Also cut a notch 3½ inches wide and ¾ inch deep from the bottom of each stile. When you assemble the rails and stiles, align the two halves of each rail face-to-face *(right)* and attach it to the stiles with screws.

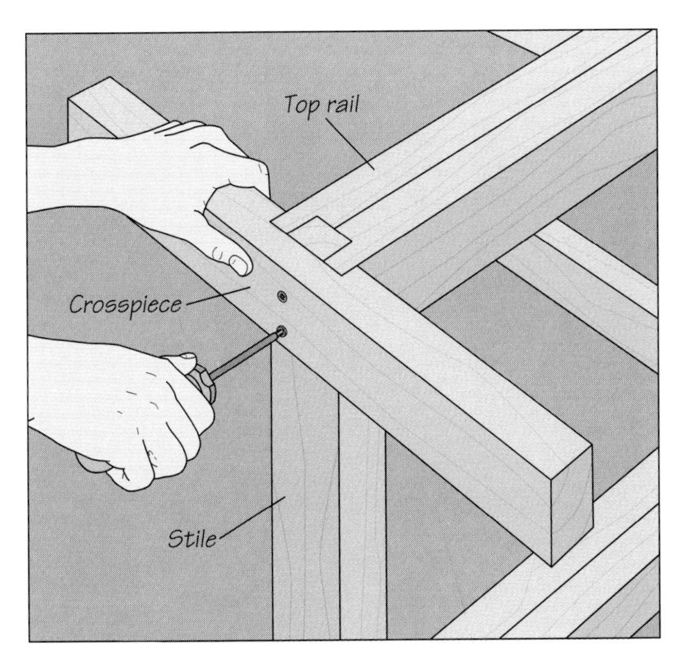

3 Mounting the crosspieces to the stiles
To join the crosspieces to the rack, cut a 3½-inch-wide dado in the middle of each piece and screw them in place *(above)*. The middle crosspiece will rest on the median rail. The top pieces will rest on the outside shoulders of the notched top of each stile.

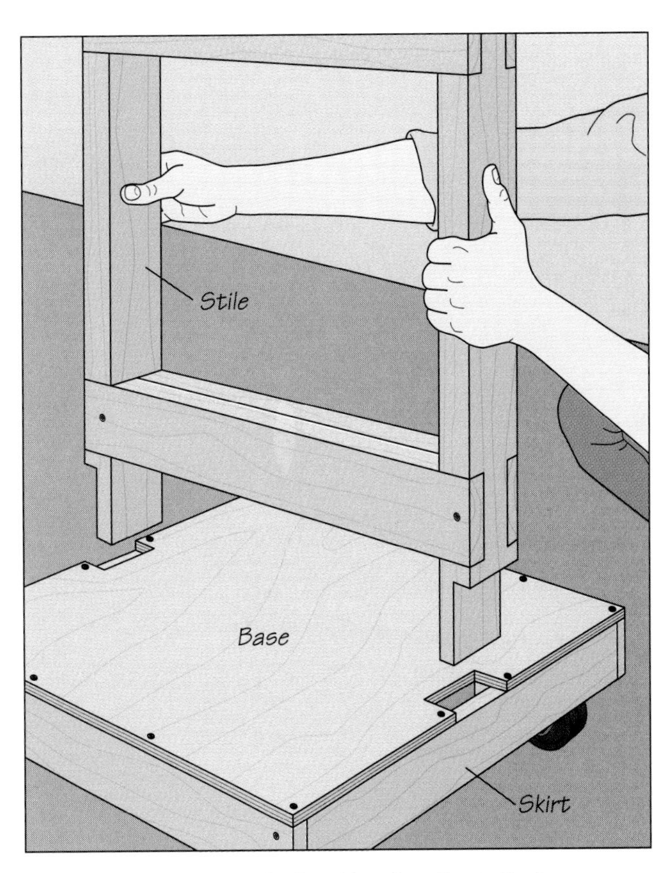

4 Attaching the stiles to the base
Finish the rack by sawing two notches in the base and skirt to accommodate the stiles, rabbeting one end of each skirt piece, and screwing them together to form a box *(page 118, inset)*. Use screws to attach the base to the skirt. Finally, attach casters to the underside of the skirt at each corner of the rack, then slip the stiles into the notches in the base *(above)* and secure the stiles to the base and skirt with glue and screws.

SHOP TIP

Storing clamps in a can
A trash can fitted with a shop-made lid serves as a convenient way to store small bar or pipe clamps. Cut a piece of ½-inch plywood into a circle slightly smaller than the diameter of the can's rim. Then scribe a series of concentric circles on the plywood to help you locate the holes for the clamp bars. Space the circles about 3 inches apart and mark points every 3 inches along them. Bore a 1-inch-diameter hole through each point, fit the piece of plywood in the can, and drop the clamps through the holes. To prevent the clamps from corroding, keep the can dry at all times.

Commercial lumber racks, such as these cantilevered lumber shelves, are both adaptable and strong, making them ideal for the home workshop. Screwed to a concrete wall or to wall studs, they can be adjusted to various heights to suit your particular storage needs.

A VERTICAL PLYWOOD RACK

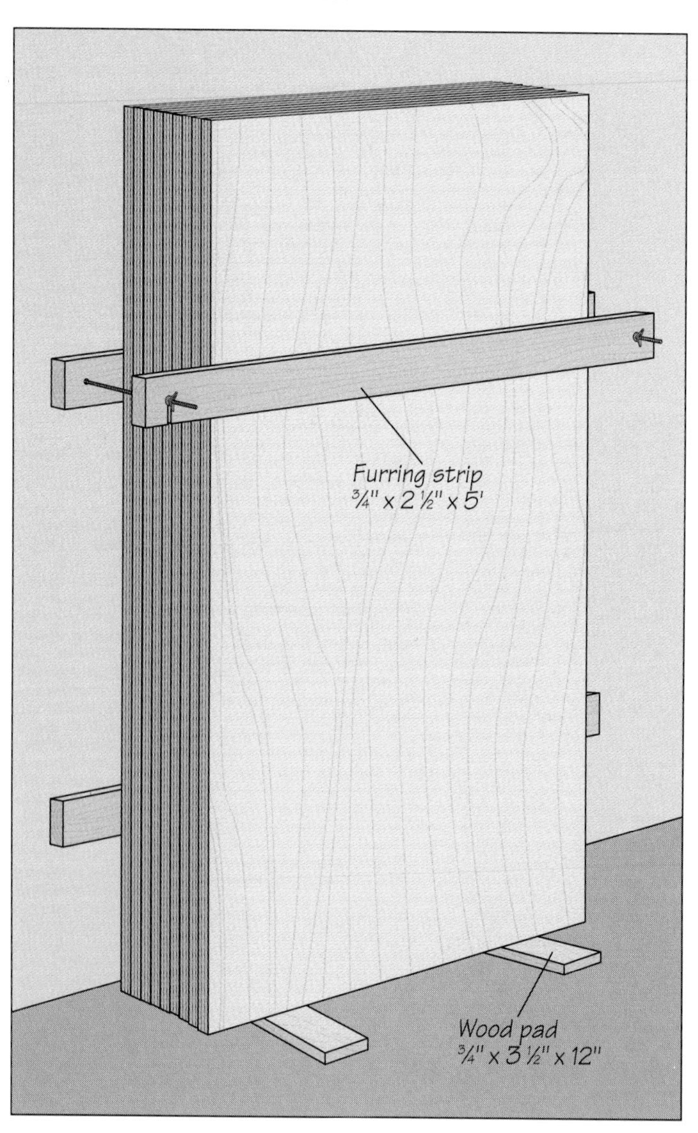

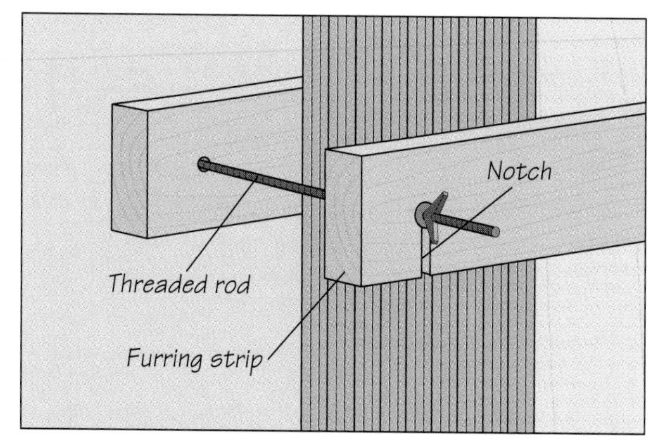

Constructing the rack

For long-term storage, stacking plywood on end saves valuable shop floor space. The rack shown at left is built from furring strips, threaded rods, and wing nuts. Start by screwing two 1-by-3 furring strips to the studs of one wall, 2 and 5 feet from the floor; first bolt two threaded rods 4½ feet apart into the top strip. Cut a third furring strip and bore a hole through it at one end and saw a notch at the other end to line up with the rods. Both openings should be slightly larger than the diameter of the rods. Place two wood pads on the floor between the rods and stack the plywood sheets upright on them. Holding the third furring strip across the face of the last panel, slip one rod through the hole and the other into the slot. Put washers and wing nuts on the rods and tighten them, pulling the furring strip tightly against the plywood *(above)*. To remove a sheet from the stack, loosen the wing nuts and swing the furring strip up and out of the way.

A LUMBER-AND-PLYWOOD RACK

Fastening the rack to an unfinished wall

The rack shown below, made entirely of 2-by-4 stock, is attached to wall studs and ceiling joists. Lumber can be piled on the arms, while plywood is stacked on edge against the support brackets. You will need at least 8½ feet of free space at one end of the rack to be able to slide in plywood panels. Begin by cutting the triangular-shaped brackets and screwing them to the studs *(right)*. Cut the footings, slip them under the brackets and nail them to the shop floor. Next, saw the uprights to length and toe-nail their ends to the footings and the joists. Cut as many arms as you need, aligning the first row with the tapered end of the support brackets. Use carriage bolts to fasten the arms to the studs and uprights, making sure the arms in the same row are level. The rack in the illustration features arms spaced at 18-inch intervals.

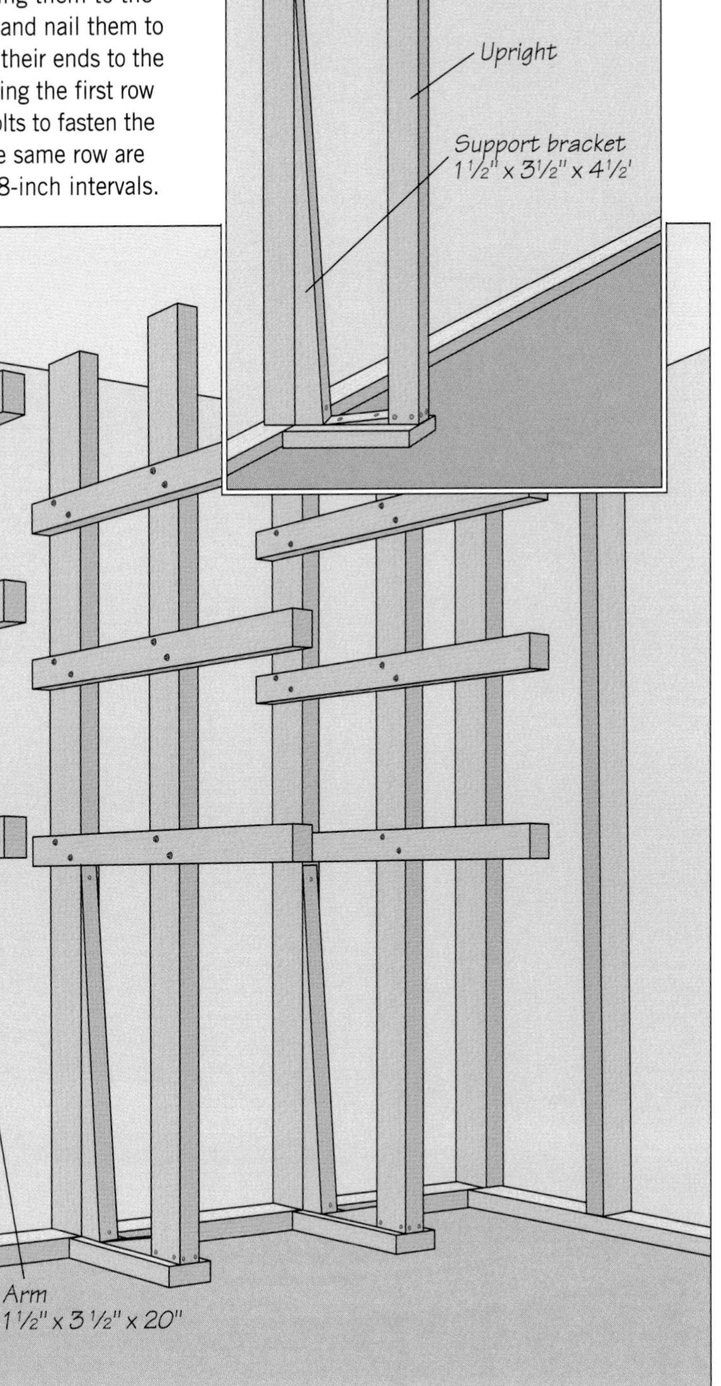

Preparing Material for Spraying
Acrylic Lacquers and Enamel
Water and Oil Based Paints
Toll-Free Hotline
1-800-626-4401
CAMPBELL HAUSFELD
PROFESSIONAL

SHOP AIDS

Most jigs that hang from the walls of woodworkers' shops typically provide a shortcut to a common task, from boring mortises to edge-gluing panels. As the previous chapters have shown, the most popular jigs are those that make a job easier and more accurate, or improve a tool's performance. But even the most mundane of workshop chores can benefit from a helping hand, whether you are moving large sheet materials around a shop or throwing some light on your work.

This chapter covers a collection of such shop aids. Some devices, such as featherboards and push sticks *(page 125)*, are indispensable for every woodworking shop. Others are designed for more specialized tasks, such as measuring and marking large circles *(page 133)* or preparing thin or small stock *(page 130)*. If you frequently work alone with large sheet materials such as plywood and particleboard, collapsible sawhorses *(page 128)* are as handy as an extra pair of hands, while an auxiliary overhead switch for the table saw *(page 131)* will make your shop a safer place. Even many of the sticky problems of finishing can be solved with a few simple devices *(page 135)*.

A spraying turntable allows you to apply a finish evenly without touching the workpiece or moving around it; the end table shown at left rests on four drying supports. As shown on page 136, the jig is easily built from plywood and a "lazy Susan" bearing.

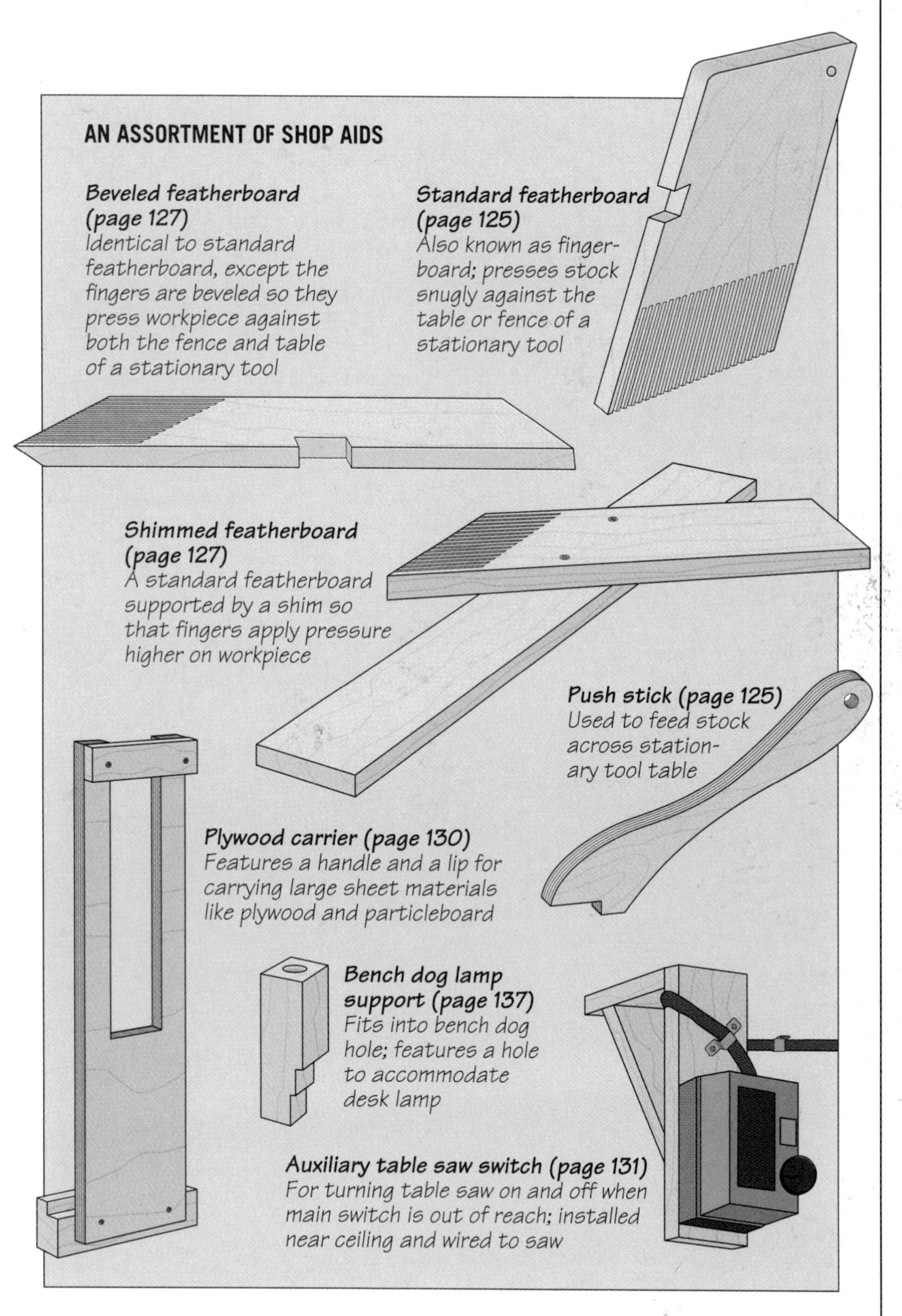

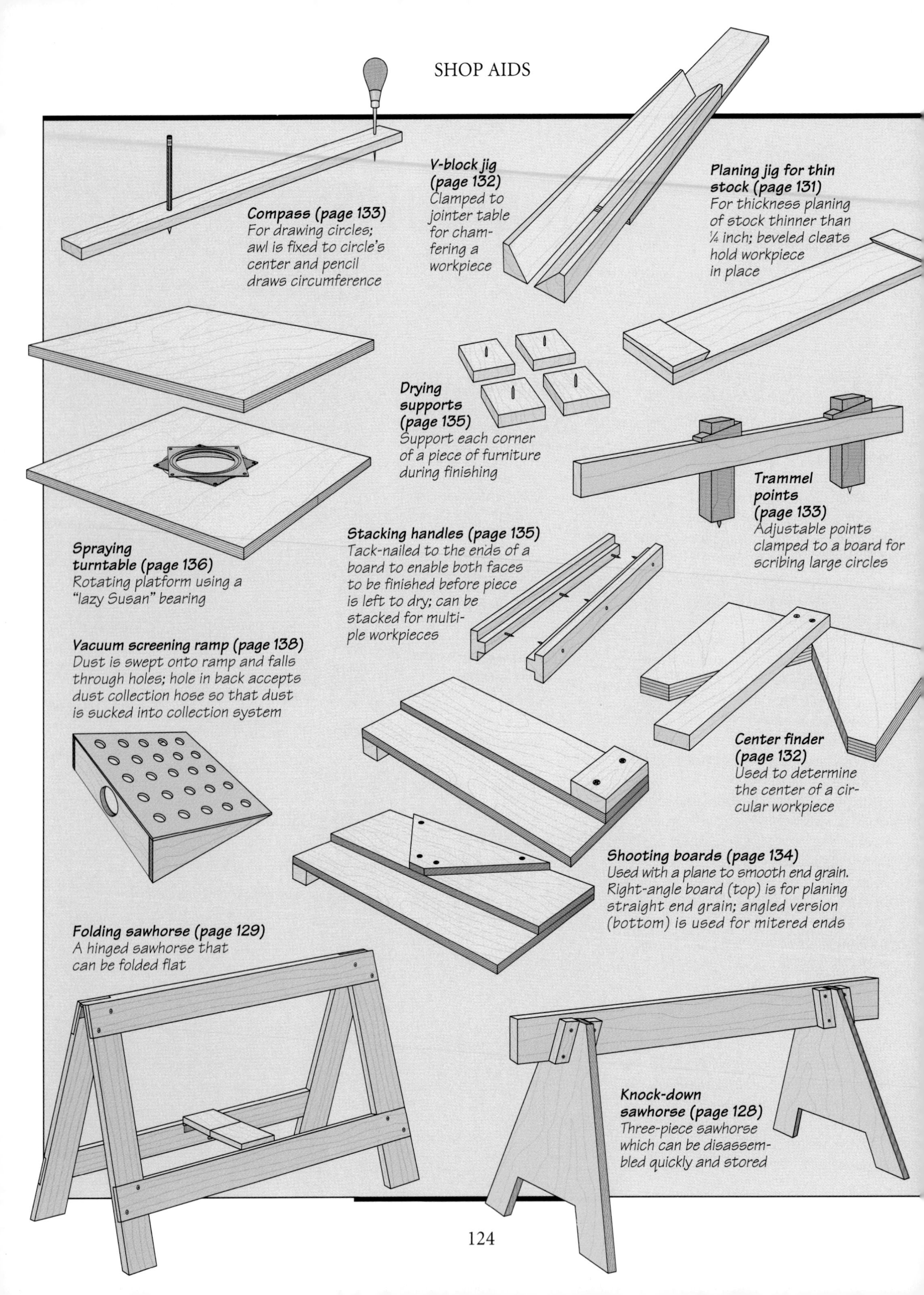
Compass (page 133)
For drawing circles; awl is fixed to circle's center and pencil draws circumference
V-block jig (page 132)
Clamped to jointer table for chamfering a workpiece
Planing jig for thin stock (page 131)
For thickness planing of stock thinner than ¼ inch; beveled cleats hold workpiece in place
Drying supports (page 135)
Support each corner of a piece of furniture during finishing
Trammel points (page 133)
Adjustable points clamped to a board for scribing large circles
Spraying turntable (page 136)
Rotating platform using a "lazy Susan" bearing
Stacking handles (page 135)
Tack-nailed to the ends of a board to enable both faces to be finished before piece is left to dry; can be stacked for multiple workpieces
Vacuum screening ramp (page 138)
Dust is swept onto ramp and falls through holes; hole in back accepts dust collection hose so that dust is sucked into collection system
Center finder (page 132)
Used to determine the center of a circular workpiece
Shooting boards (page 134)
Used with a plane to smooth end grain. Right-angle board (top) is for planing straight end grain; angled version (bottom) is used for mitered ends
Folding sawhorse (page 129)
A hinged sawhorse that can be folded flat
Knock-down sawhorse (page 128)
Three-piece sawhorse which can be disassembled quickly and stored

PUSH STICKS AND PUSH BLOCKS

Making push sticks and push blocks
Push sticks and push blocks for feeding stock across the table of a stationary power tool can be made using ¾-inch plywood or solid stock. No one shape is ideal; a well-designed push stick should be comfortable to use and suitable for the machine and task at hand. For most cuts on a table saw, design a push stick with a 45° angle between the handle and the base *(right, top)*. Reduce the handle angle for use with the radial arm saw. The notch on the bottom edge must be deep enough to support the workpiece, but shallow enough not to contact the saw table. The long base of a rectangular push stick *(right, middle)* enables you to apply downward pressure on a workpiece. For surfacing the face of a board on a jointer, the long, wide base of a push block *(right, bottom)* is ideal. It features a lip glued to the underside of the base, flush with one end. Screw the handle to the top, positioning it so the back is even with the end of the base.

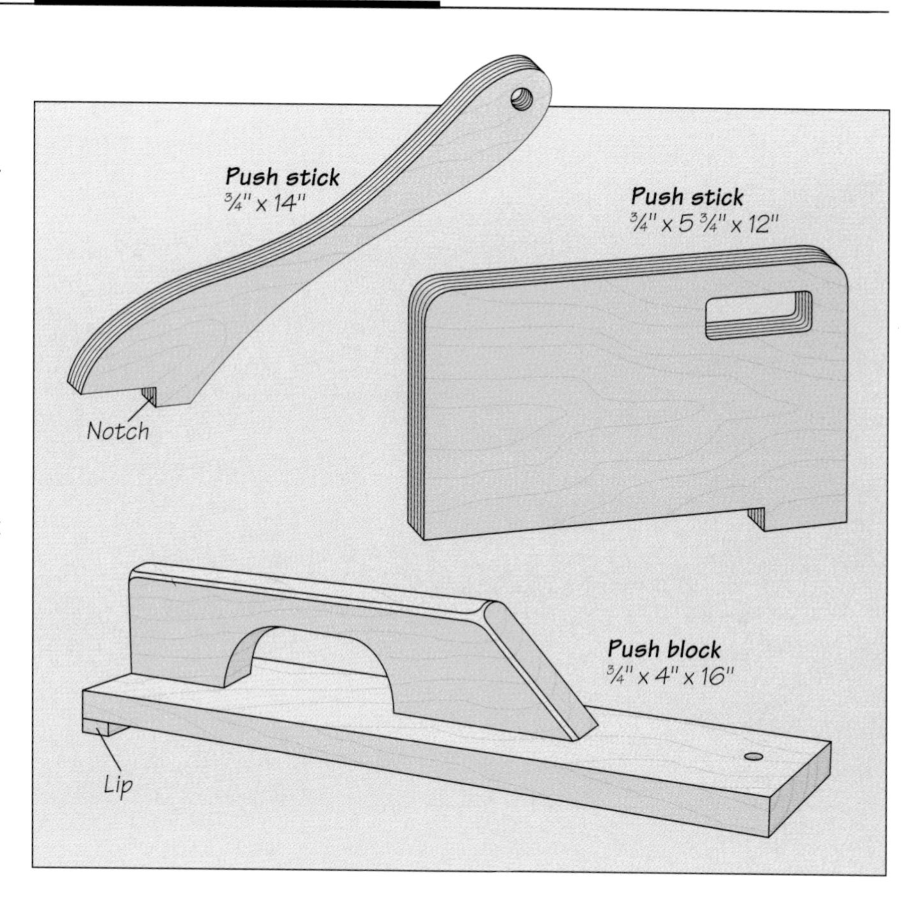

A STANDARD FEATHERBOARD

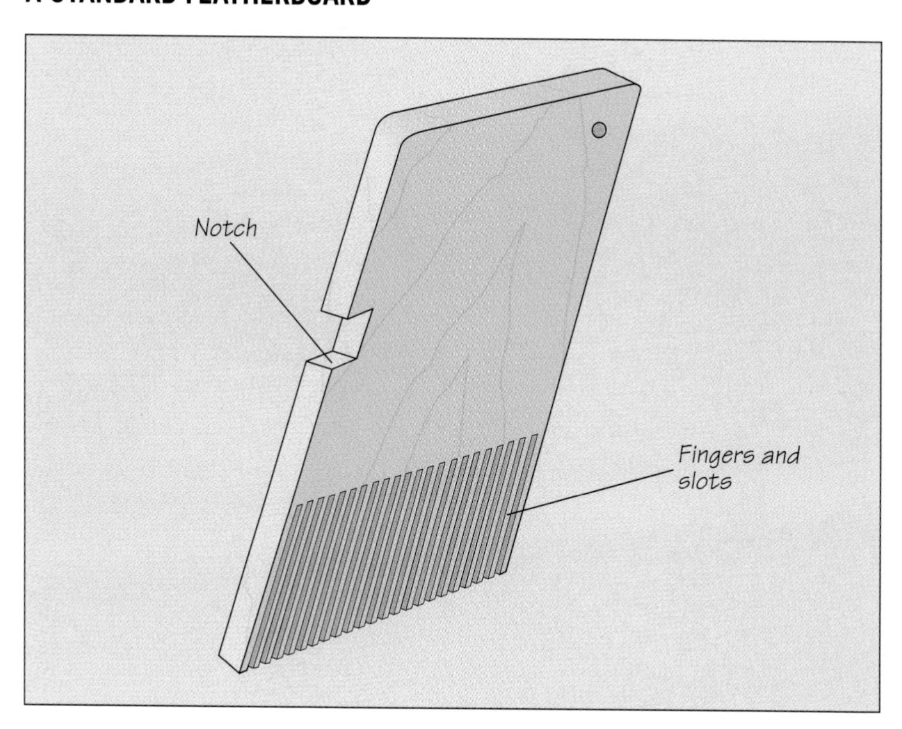

1 Making a standard featherboard
Featherboards serve as anti-kickback devices, since the fingers allow the workpiece to move in only one direction—toward a stationary tool's bit or blade. To make a featherboard like the one shown at left, cut a 30° to 45° miter at one end of a ¾-inch-thick, 3- to 4-inch-wide board; the length of the jig can be varied to suit the work you plan to do. Mark a parallel line about 5 inches from the mitered end and cut a series of slots to the marked line on the band saw, spacing the kerfs about ⅛ inch apart to create a row of sturdy but pliable fingers. Finally, cut a notch out of one edge of the featherboard to accommodate a support board.

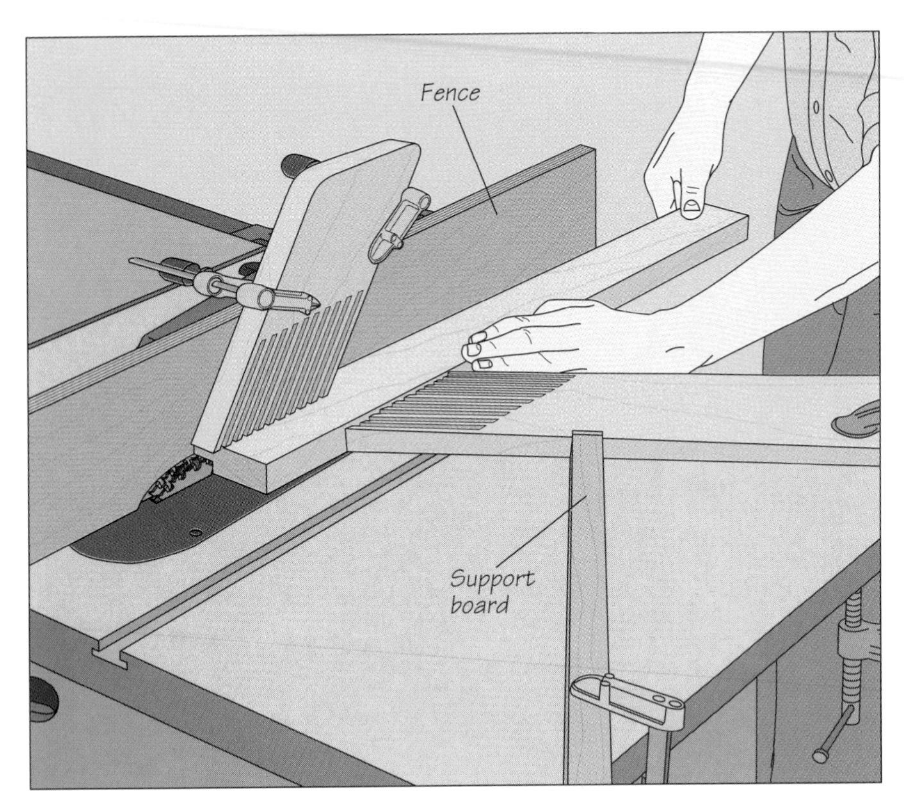

2 Using standard featherboards on the table saw

Clamp one featherboard to the fence above the blade, and place a longer one halfway between the blade and the front of the table. Clamp a support board in the notch perpendicular to the horizontal featherboard to prevent it from creeping out of place during the cut. For the operation shown at left, feed the workpiece into the blade until your trailing fingers reach the featherboards. Then use a push stick to finish the cut, or move to the back of the table with the saw still running and pull the workpiece past the blade.

Push sticks and featherboards make an operation like ripping on the table saw much safer by keeping your hands well away from the blade. The push stick is used to feed the stock and keep it flat on the table, while the featherboard presses the workpiece against the fence. The featherboard shown in the photo is secured to the table with special hardware rather than with clamps. A clamping bar in the miter slot features two screws that can be tightened, causing the bar to expand and lock tightly in the slot.

A BEVELED FEATHERBOARD

Ripping stock with a beveled featherboard
A featherboard clamped to the fence of a table saw, as shown on the previous page, can get in the way of a push stick during a rip cut. A featherboard with a beveled end will press a workpiece against both the fence and saw table, eliminating the need to clamp a featherboard to the fence *(right)*. Make the device as you would a standard featherboard *(page 125)*, but cut a 45° bevel on its leading end before cutting the fingers and slots. Also make sure that the featherboard is thicker than the stock you are ripping *(inset)*.

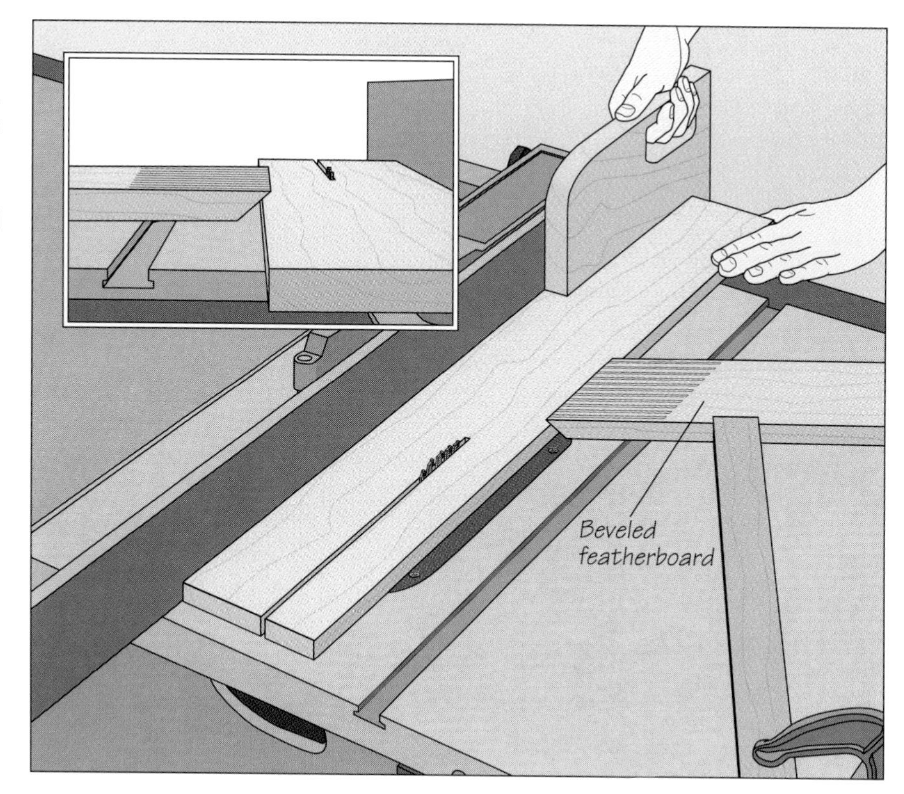

A SHIMMED FEATHERBOARD

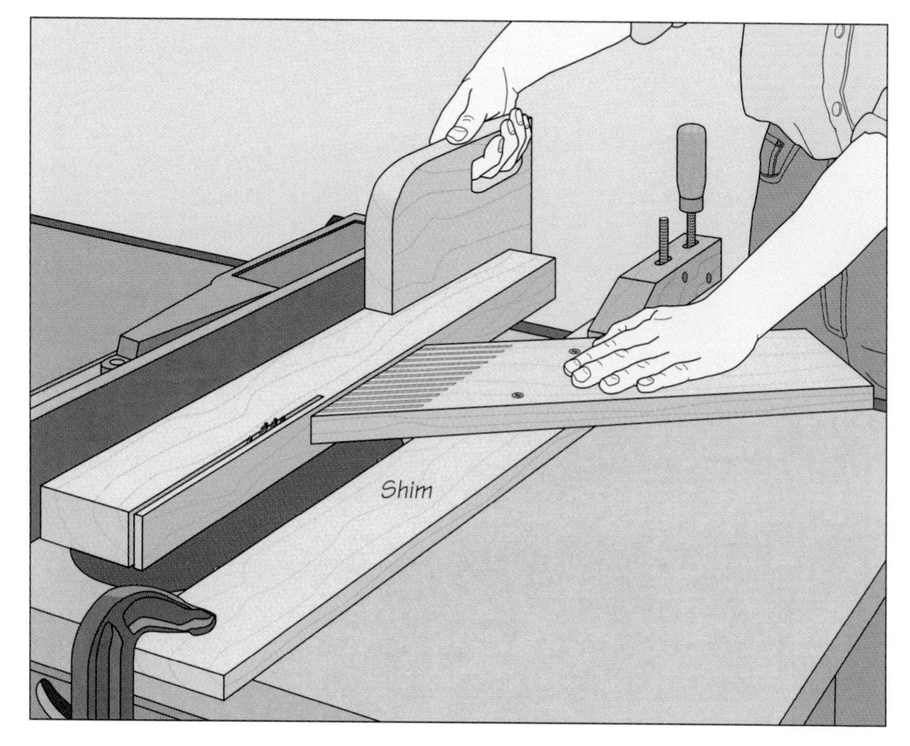

Shimming a featherboard
When working with thick stock or running a board on edge across a saw table, a featherboard clamped directly to the table may apply pressure too low on the workpiece, causing it to tilt away from the fence. To apply pressure closer to the middle of the stock, screw the featherboard to a shim and then clamp the shim to the table *(left)*.

A KNOCKDOWN SAWHORSE

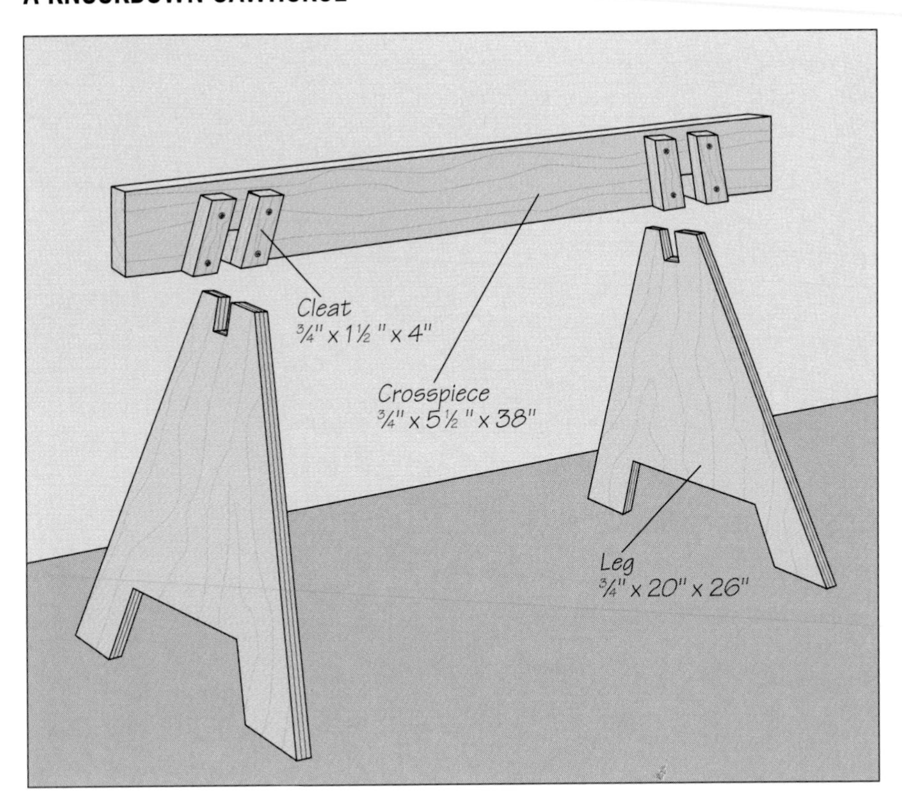

Building the sawhorse

A sturdy sawhorse like the one shown at left can be made with only a small amount of lumber and plywood, and taken apart and reassembled quickly. Cut the legs from ¾-inch plywood, then saw a 3-inch-deep notch in the middle of the top of both pieces. Next, cut the crosspiece from 1-by-6 stock and saw a 1½-inch-deep slot 8 inches in from either end to fit into the legs. Angle the slots roughly 5° from the vertical to spread the legs slightly outward. For added stability, screw 4-inch-long 1-by-2 cleats to the crosspiece on each side of the slots.

To plane long workpieces without the help of a bench vise, clamp them edge-up with handscrews on two or more sawhorses, as shown at right. Secure the handscrews to the crosspieces with C clamps to prevent them from slipping.

A FOLDING SAWHORSE

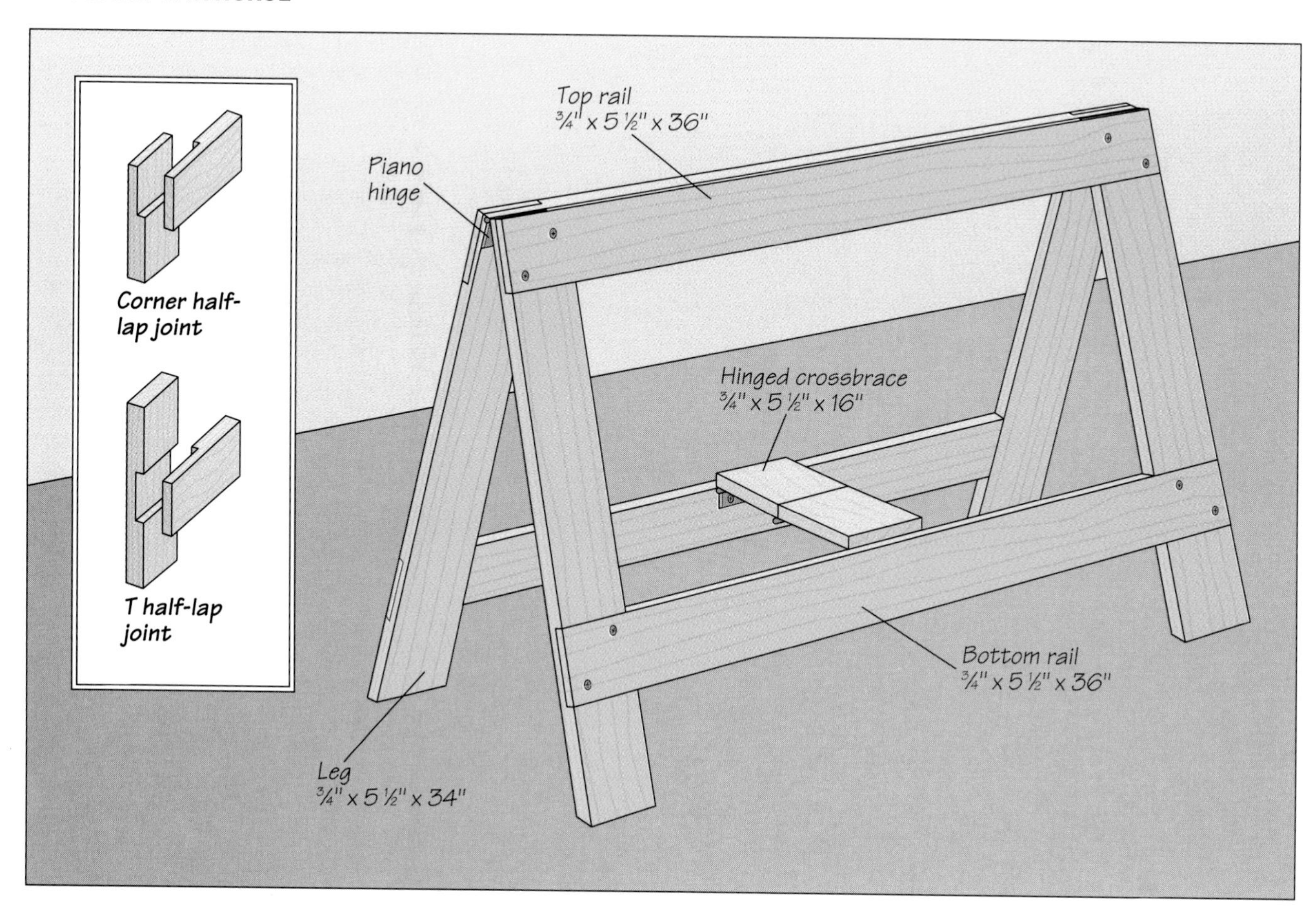

SHOP TIP

Padding sawhorses

To prevent a sawhorse from marring your work, cover its crosspiece with a strip of carpet, an old towel, or a blanket. Fold the material over the top edge of the crosspiece and screw or nail it to the sides.

Making the sawhorse

Made entirely from 1-by-6 stock, the lightweight sawhorse shown above, featuring a hinged top and crossbrace, folds flat for easy storage. Cut the legs and rails to length and join them together; use corner half-laps *(inset, top)* to join the top rails to the legs and T half-laps *(inset, bottom)* to attach the legs to the bottom rails. Reinforce the joints with glue and screws, then join the two sections at the top rails with a piano hinge. Finally, cut the crossbrace; be sure it is long enough so that when the legs are spread, the piano hinge is recessed below the top edge of the top rails. Saw the crossbrace in half and connect the two pieces with the hinge. Then fasten the crossbrace to both side rails, again using piano hinges.

A PLYWOOD CARRIER

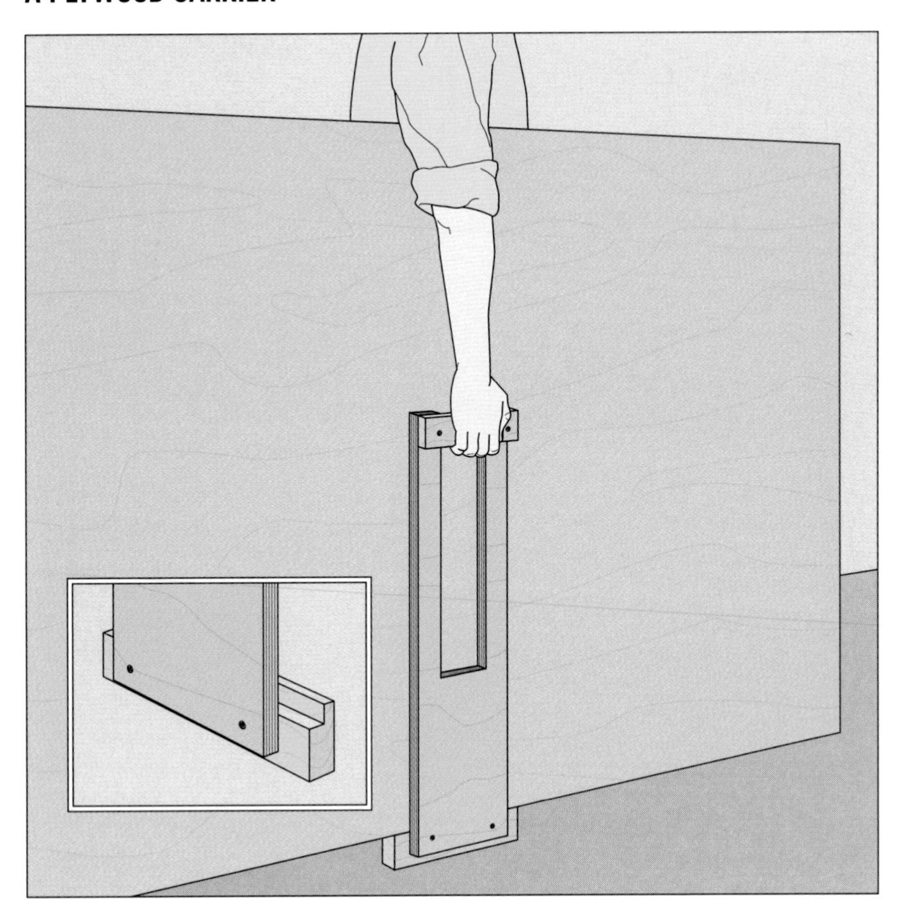

Moving large panels
Sheets of plywood, particleboard, and hardboard are often heavy and awkward to carry. The carrier shown at left will make the load easier to bear. Rout a 1-inch-wide rabbet along one edge of a 12-inch-long board. Cut a notch out of one end of a piece of plywood, then screw a wood block across the end of the notch to serve as a handle. Attach the other end of the plywood piece to the rabbeted face of the board *(inset)*. To use the carrier, simply hook it under the sheet and pull it up under your arm *(left)*. Some woodworkers find it more comfortable to stand on the carrier side of the panel and use their other hand to steady it.

PLANING SHORT AND THIN STOCK

Using runner guides to plane short stock
Feeding short boards through a thickness planer can cause sniping and kickback. To hold short stock steady as it enters and exits the planer, glue two solid wood scrap runners to the edges of your workpiece. Make sure the runners are the same thickness as the workpiece and extend several inches beyond both ends. Feed the workpiece into the planer *(right)*, making a series of light cuts until you have reached the desired thickness. Then cut off the runners.

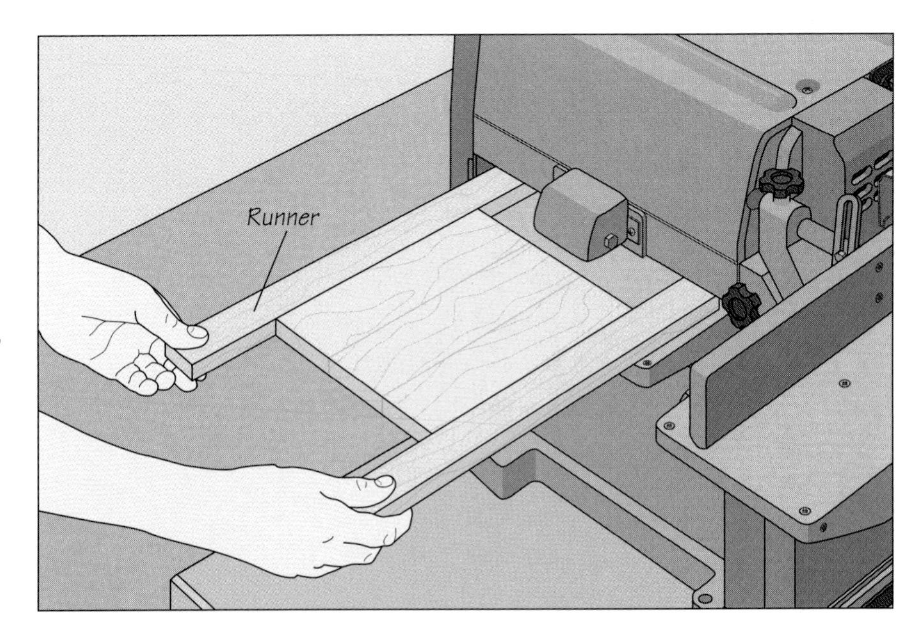

Using a planing jig for thin stock
Thickness planing stock thinner than ¼ inch often causes chatter and splintering of the workpiece. To avoid these problems make thin stock "thicker" with this jig. To make it, simply glue two beveled cleats to either end of a board that is slightly longer than your workpiece *(inset)*. To make the cleats, cut a 45° bevel across the middle of a board approximately the same thickness as the workpiece. Next, bevel the ends of the workpiece. Set the stock on a backup board, position the cleats flush against the workpiece so the bevel cuts are in contact, and glue the cleats in place to the backup board. Run the jig and workpiece through the planer, making several light passes down to the desired thickness *(right)*, then crosscut the ends of the workpiece square.

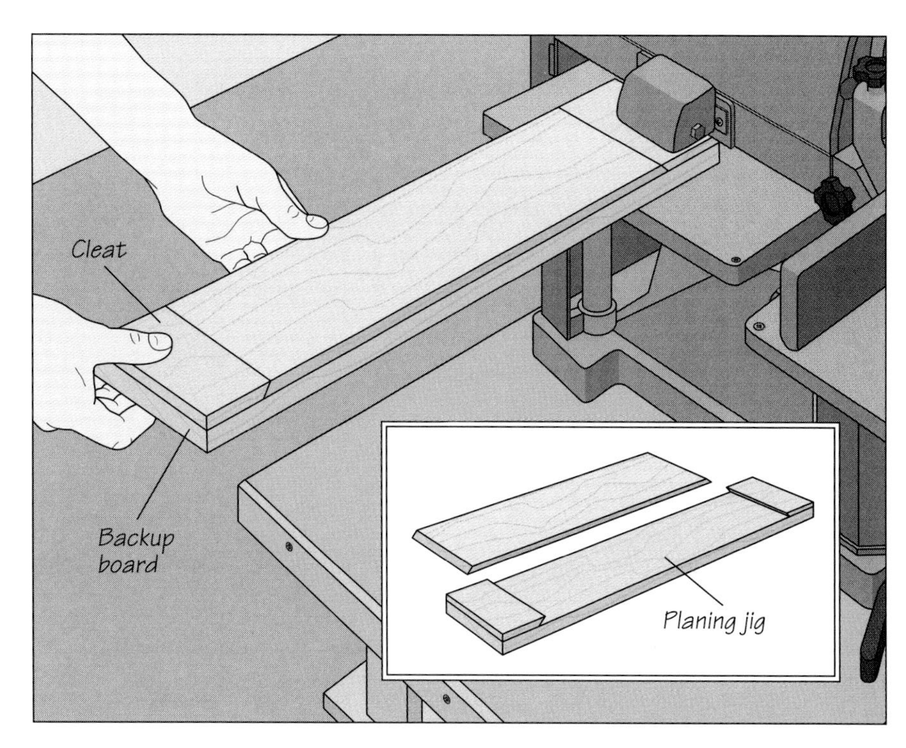

AN AUXILIARY SWITCH FOR THE TABLE SAW

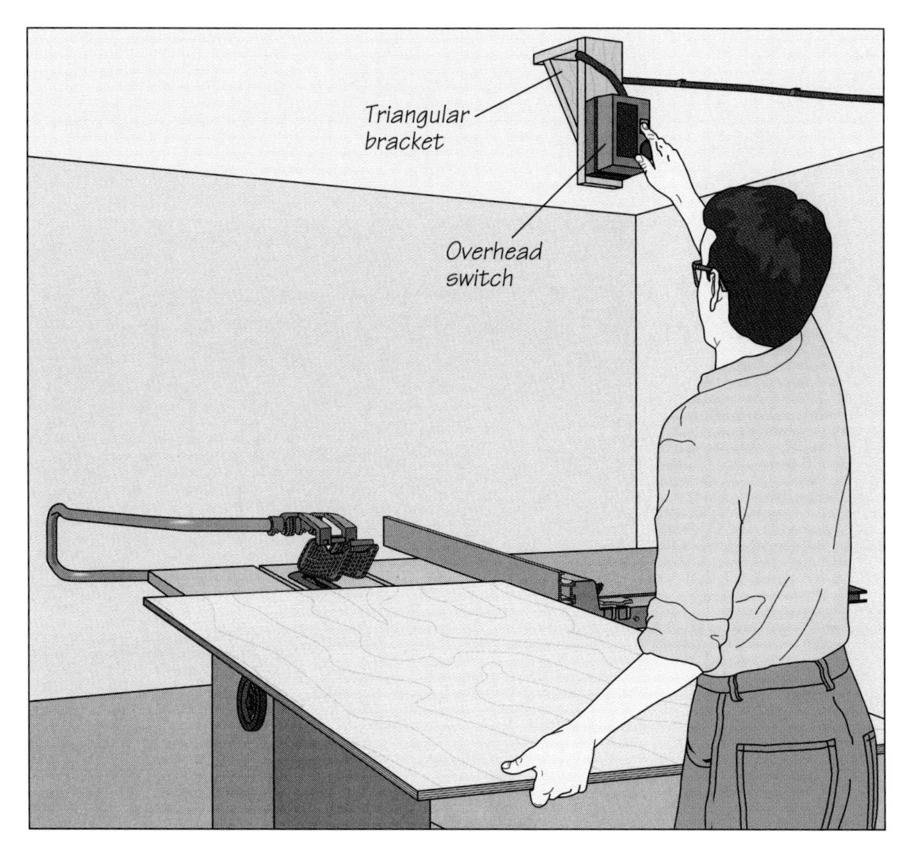

Installing an overhead switch
Switching on a table saw while balancing a large panel on the table can prove difficult. The addition of an overhead switch will enable you to start the saw when the main switch is out of reach *(left)*. Locate the new switch so you can reach it comfortably with a 4-by-8 panel on the saw table just in front of the blade; screw a triangular bracket to the ceiling and attach the switch to the bracket at a suitable height. Run a length of non-metallic sheathed 12-gauge cable from the switch along the ceiling, down the wall, and across the floor to your saw. Have a licensed electrician wire the switch to the saw so that both it and the original switch are able to start or stop the machine; **never disconnect the switch on the saw itself.**

V-BLOCK JIG

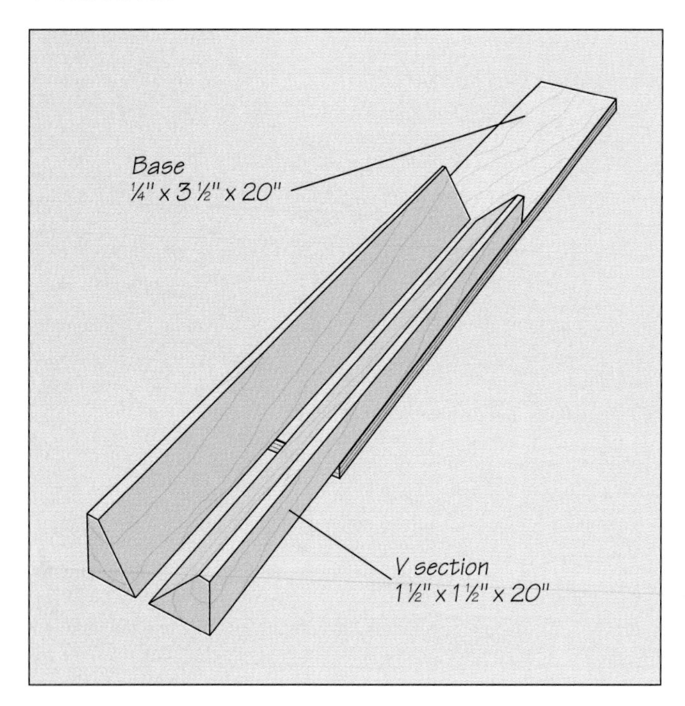

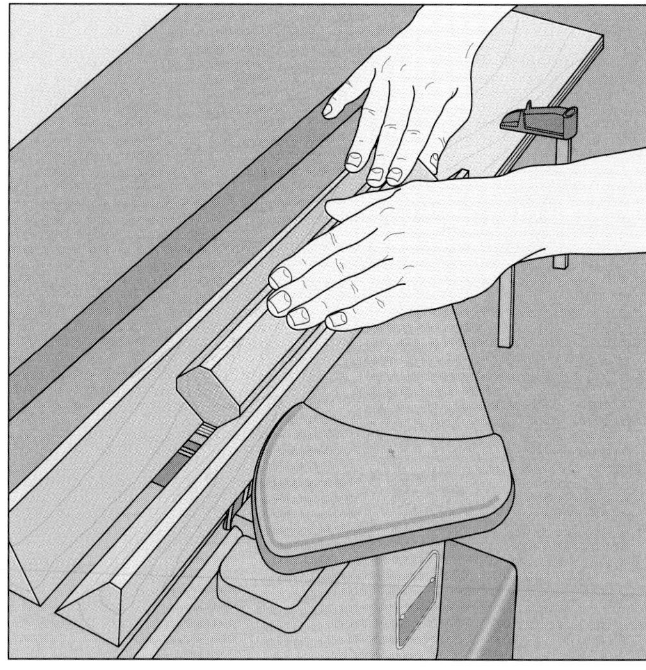

Cutting a chamfer
To cut chamfers on your jointer, use the simple jig shown above. Refer to the illustration for suggested dimensions. Begin by bevel cutting 2-by-2s for the V section of the jig. Position the two cut pieces on the base so they extend beyond one end by about 12 inches, with a ½-inch gap between them. Attach the two pieces to the base with countersunk screws to avoid scratching the jointer table. To use the jig, clamp it in place with one end of the base aligned with the cutterhead-end of the infeed table. Lower the infeed table until the V section of the jig lies flush on the jointer's outfeed table. Seat the workpiece in the gap of the jig, then feed it across the knives while holding it firmly in the V *(above)*.

Finding the center of a circular workpiece is easy if you use the jig shown at right. The simple device consists of a piece of plywood with a 90° wedge cut out of it and a 12-inch-long 1-by-2 mounted so that one edge bisects the wedge. To use the jig, seat the workpiece in the wedge and draw a line across its diameter using the 1-by-2 as a guide. Rotate the workpiece about 90° and draw another line. The two will intersect at the center of the circle.

TWO DEVICES FOR SCRIBING CIRCLES

Making and using trammel blocks
You can scribe a large circle using a set of trammel blocks like the one shown at right. Cut the pieces of the jig from solid wood, referring to the inset for dimensions; make sure the beam is longer than the radius of your circle. Cut an angled notch ½ inch from the top of each block to accommodate the beam and a wedge. For the pivot, drive a nail into the bottom of one block, snip off the head, and file it to a point. Mount a sharp pencil in a hole bored in the bottom of the other block. Make sure its point is level with the nail. To use the jig, loosen the wedges and slide the blocks along the beam until the gap between the nail tip and pencil point equals the desired radius of your circle. Tighten the wedges, hold the pivot point steady at the center of the circle, and rotate the pencil point around it *(right)*.

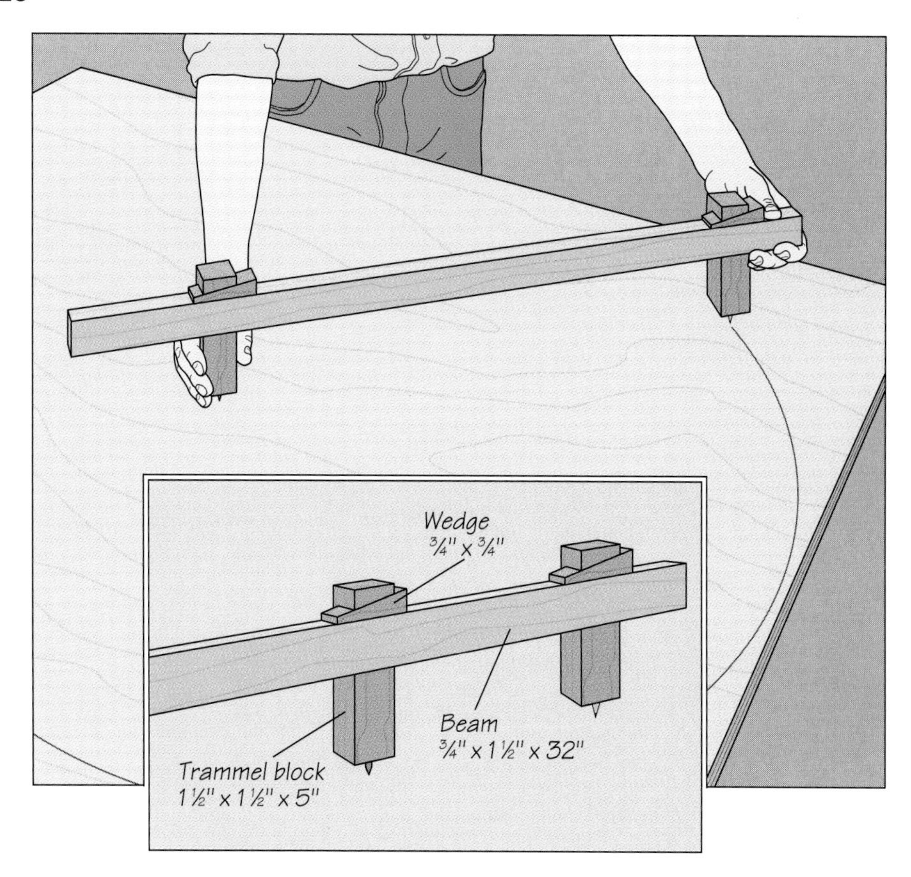

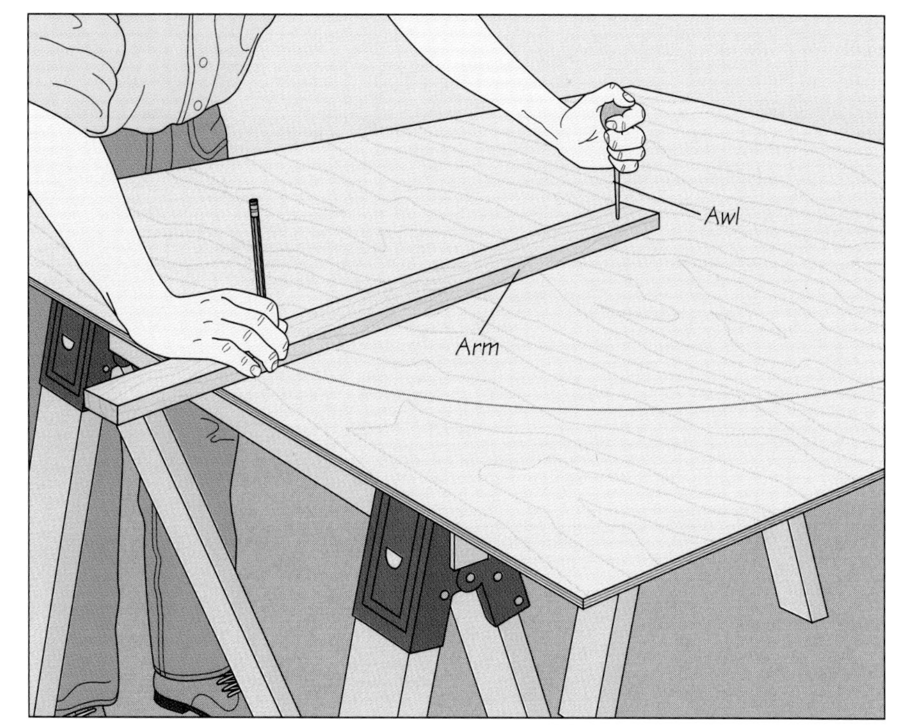

Making and using a fixed compass
Consisting of an arm, an awl, and a pencil, the compass shown at left will allow you to scribe a circle of virtually any radius. For the arm, cut a 1-by-2 a few inches longer than the radius of your circle. Bore a hole about 1 inch from one end of the arm, large enough to hold the shaft of the awl. Make another hole big enough to accommodate the pencil; the distance between the holes should equal the radius of the circle. Fit the awl and sharpened pencil into their respective holes, making sure the two extend from the bottom of the arm by the same amount. Use the compass as you would trammel blocks, holding the tip of the awl at the center of the circle and rotating the pencil around it to scribe the circle *(left)*.

TWO SHOOTING BOARDS

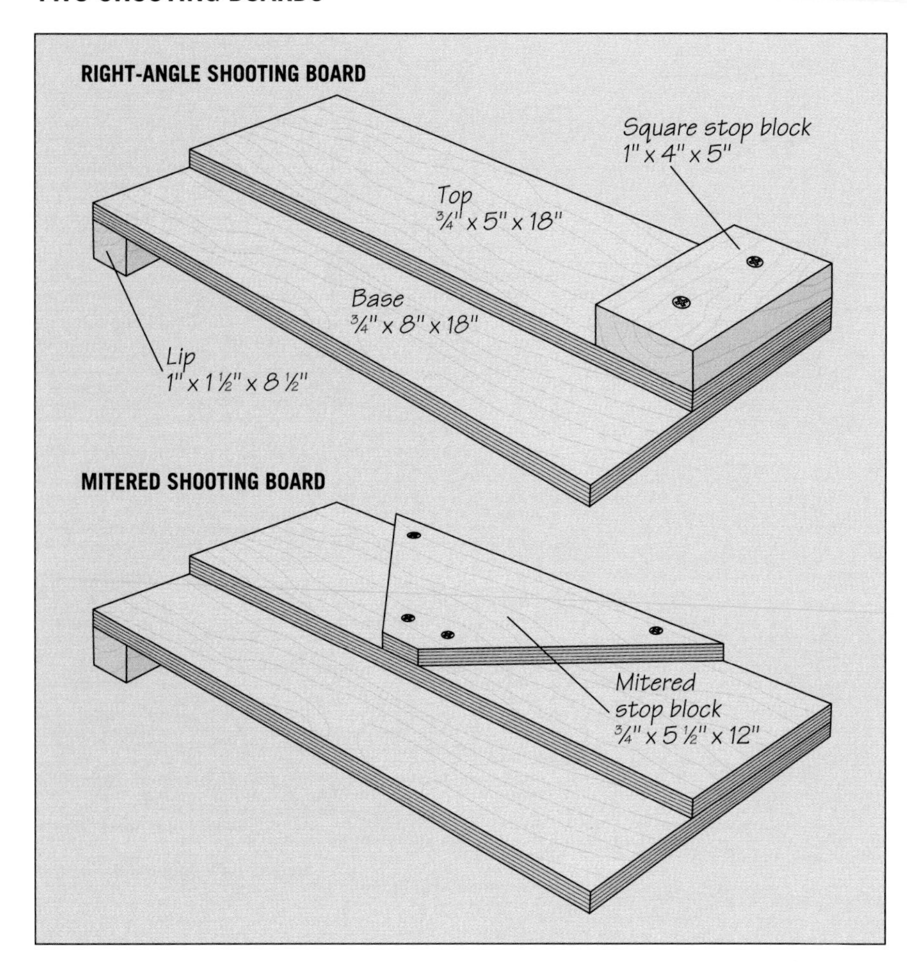

1 Making shooting boards
To smooth end grain with a plane, use a shooting board like those shown at left. The right-angle shooting board *(left, above)* is for planing straight end grain; a mitered version can also be built *(left, below)*. Cut the pieces according to the dimensions suggested in the illustrations. Build the base, top, and mitered stop block from ¾-inch plywood; use solid wood for the lip and the square stop block. Screw the top to the base with the ends and one edge aligned. Then attach the lip to the base, making sure that the lip lines up with the end of the base. For the right-angle shooting board, fasten the stop block to the top flush with the other end of the jig. For the mitered shooting board, center the stop block on the top.

2 Smoothing end grain
To use either jig, hook the lip on the edge of a work surface. Set your workpiece on the top, butting the edge against the stop block so that it extends over the edge of the top by about ¹⁄₁₆ inch. With the mitered shooting board, position the workpiece against the appropriate side of the stop block. (For a long workpiece, it may be necessary to place a support board under the opposite end to keep the workpiece level.) Set a plane on its side at one end of the jig and butt the sole against the edge of the top. Holding the workpiece firmly, guide the plane along the jig from one end to the other *(right)*.

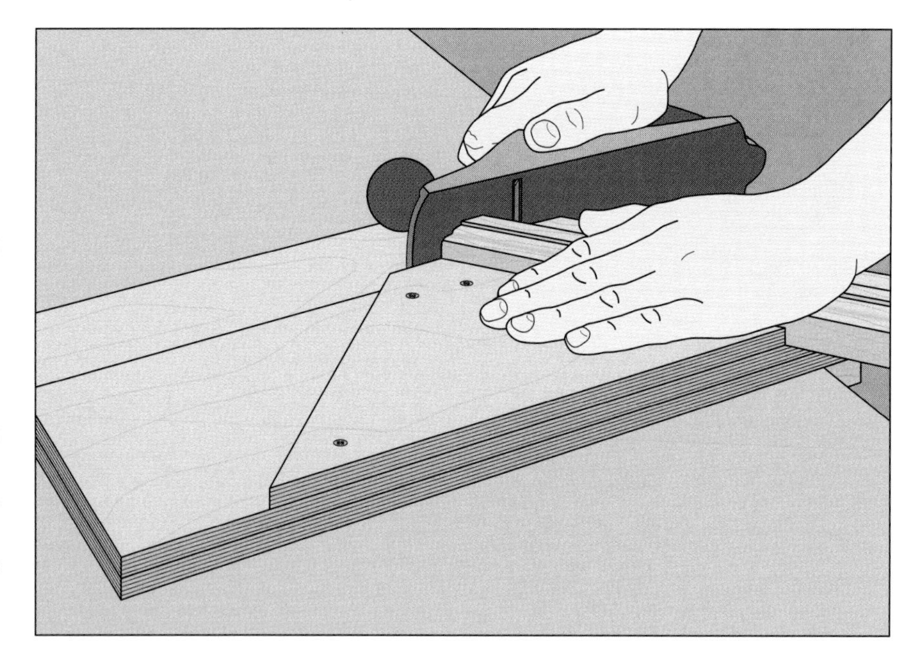

A clean metal can with wire strung across the mouth offers a neat and simple way to control the flow of stain or varnish from a brush. Punch holes in opposite sides of the can near the rim, string a wire between the holes and pour the liquid into the container. After dipping the brush, draw the bristles across the wire to wipe off any excess liquid.

STACKING HANDLES AND DRYING SUPPORTS

Stacking shelves to dry
Finishing a shelf one side at a time doubles the time needed for this task. Using the simple stacking handles shown at left, you can finish both sides at once. Cut the handles from solid wood stock and mill a tongue in one face of each one. Make the handles at least ½ inch wider than the thickness of the stock you are finishing. Drive small nails longer than the handles' thickness through them and gently press the protruding points into the ends of the workpiece before finishing it. Use the handles to turn the board as you apply finish; when you are done, the boards can be stacked, allowing air to circulate freely as the finish dries. When finishing a larger piece of furniture, set the piece on a set of drying supports *(inset)*. These 2-inch-square wood blocks have small nails driven through their centers to support a workpiece at its corners.

A SPRAYING TURNTABLE

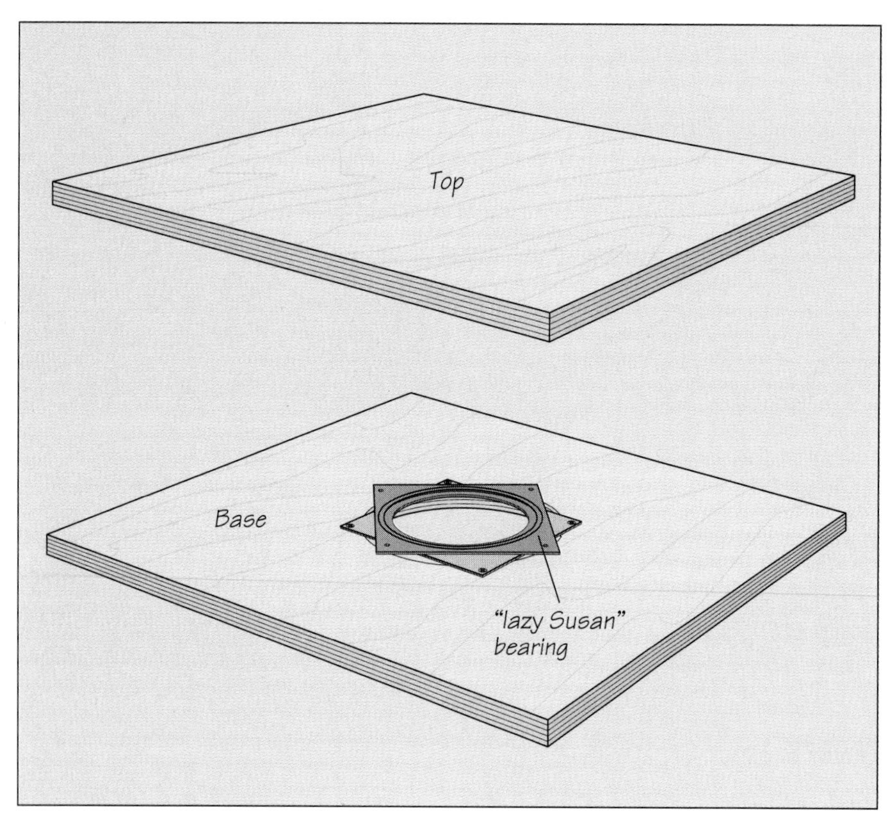

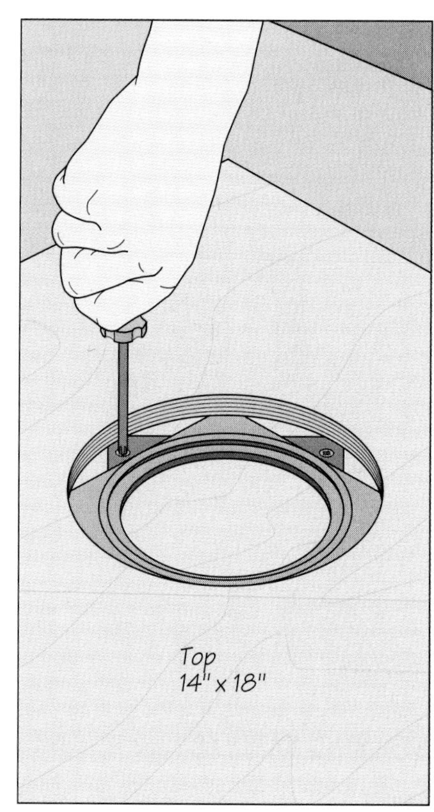

1 Building the turntable
Consisting of a base and top cut from ¾-inch plywood with a "lazy Susan" bearing fastened in between, the turntable shown above allows a piece of furniture to be rotated as it is being sprayed with a finish. Cut the base and top slightly larger than the base of the piece of furniture to be finished. Cut a hole in the center of the base to allow access to the screw holes for attaching the upper bearing to the top once the lower bearing is secured to the base. First attach the lower bearing to the base with screws. To fasten the jig top, set the base on top of it with the bearing sandwiched between the two pieces and flip them upside down. With the edges of the pieces flush, rotate the bearing so the remaining screw holes are exposed, then screw the upper bearing to the top *(above, right)*.

SHOP TIP

Lubricating tools
Lightly oiling a handsaw blade or a plane sole before storing it keeps the tool clean and prevents rust. To oil your tools neatly, use a lubricating pad made from a long strip of burlap, tightly rolled and packed in a small can as shown here. Make sure the strip is wide enough to extend past the top of the can. Soak the material in thin machine oil, and wipe your plane soles and handsaw blades over it before storing the tools. Tightly cover the can when not in use.

2 Using the turntable
Make four small drying supports *(page 135)*. Set the workpiece on the tips of the nails, then slowly rotate the turntable with one hand while operating a spray gun with the other *(above)*.

BENCH DOG LAMP SUPPORT

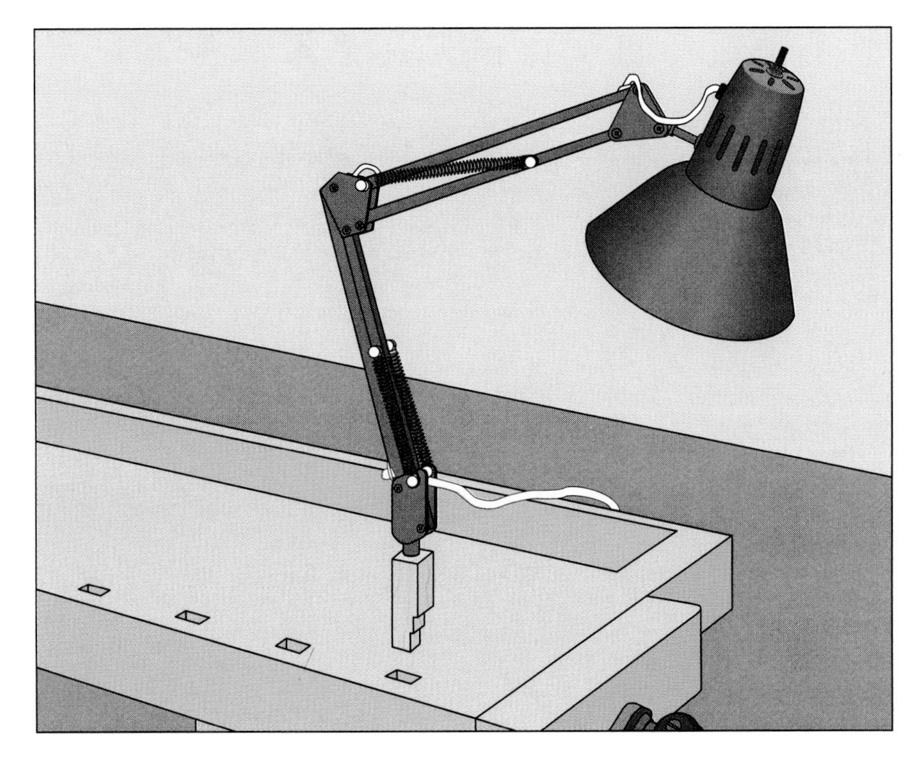

A movable light for a workbench
A desk lamp attached to a bench dog as shown at left will enable you to position the light at any of the dog holes along the bench. To make the jig, bore a hole the same diameter as the shaft of the lamp into the head of a wooden bench dog *(page 92)*.

A VACUUM SCREENING RAMP

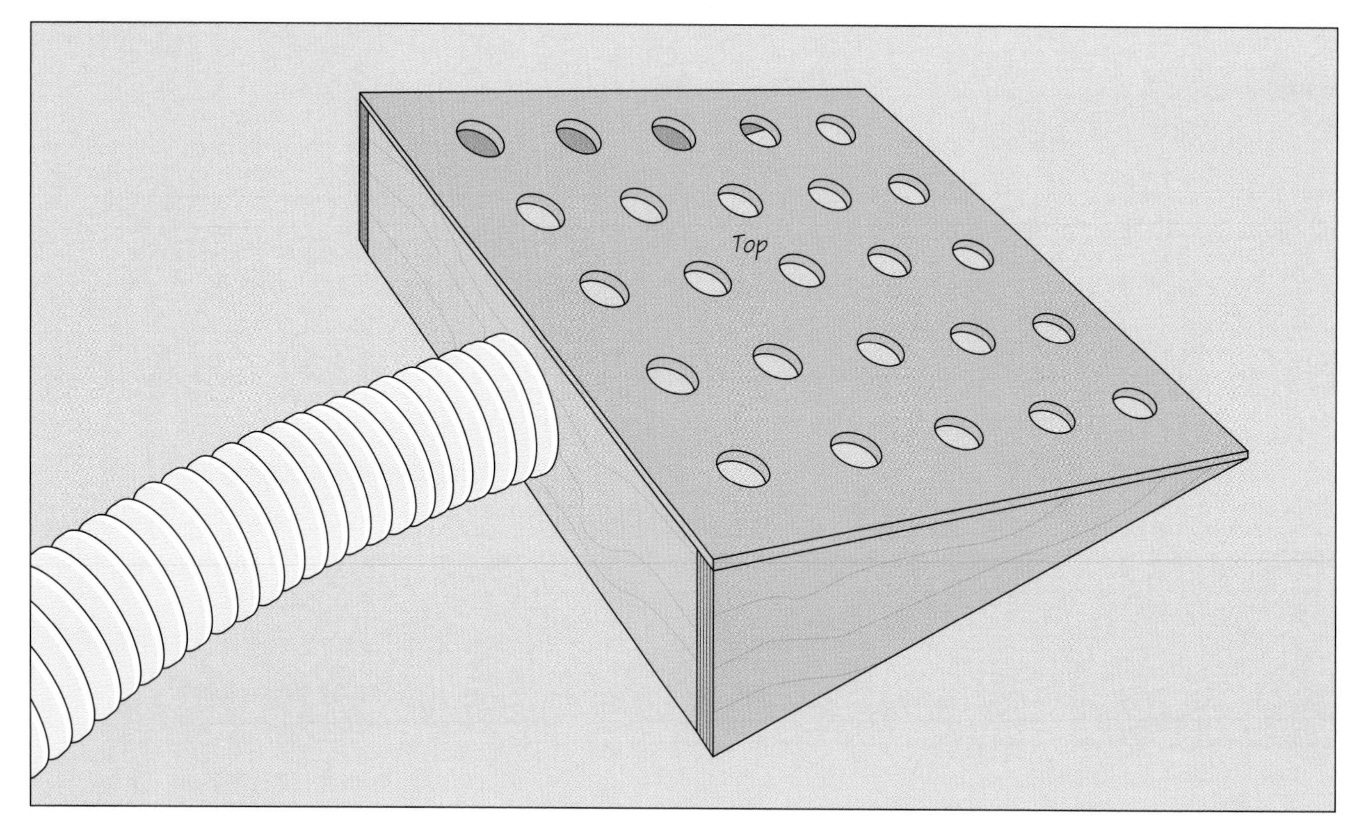

Collecting shop dust

For cleaning dust off the shop floor, build a wedge-shaped screening ramp from plywood and hardboard *(above)*. Before assembling the pieces, cut a hole in the back to fit a dust collection hose, and bore five rows of 1½-inch-diameter holes through the top. When dust and wood chips are swept up onto the ramp, smaller particles will fall through the holes and continue on to the dust collector. Larger debris will remain on the ramp for easy disposal.

SHOP TIP

A drawer-slide positioning jig

To help you correctly position commercial slides on drawer sides, use the jig shown here. Cut a rabbet in a scrap board; make the depth of the rabbet equal to the desired distance between the slide and the bottom of the drawer side. To use the jig, hold it against the bottom of the drawer side. Then set the slide on the drawer side, with the bottom edge of the hardware butted against the jig. Holding the slide and the jig in place, mark the screw holes, bore pilot holes, and screw the slide to the drawer.

A STEAM-BENDING JIG

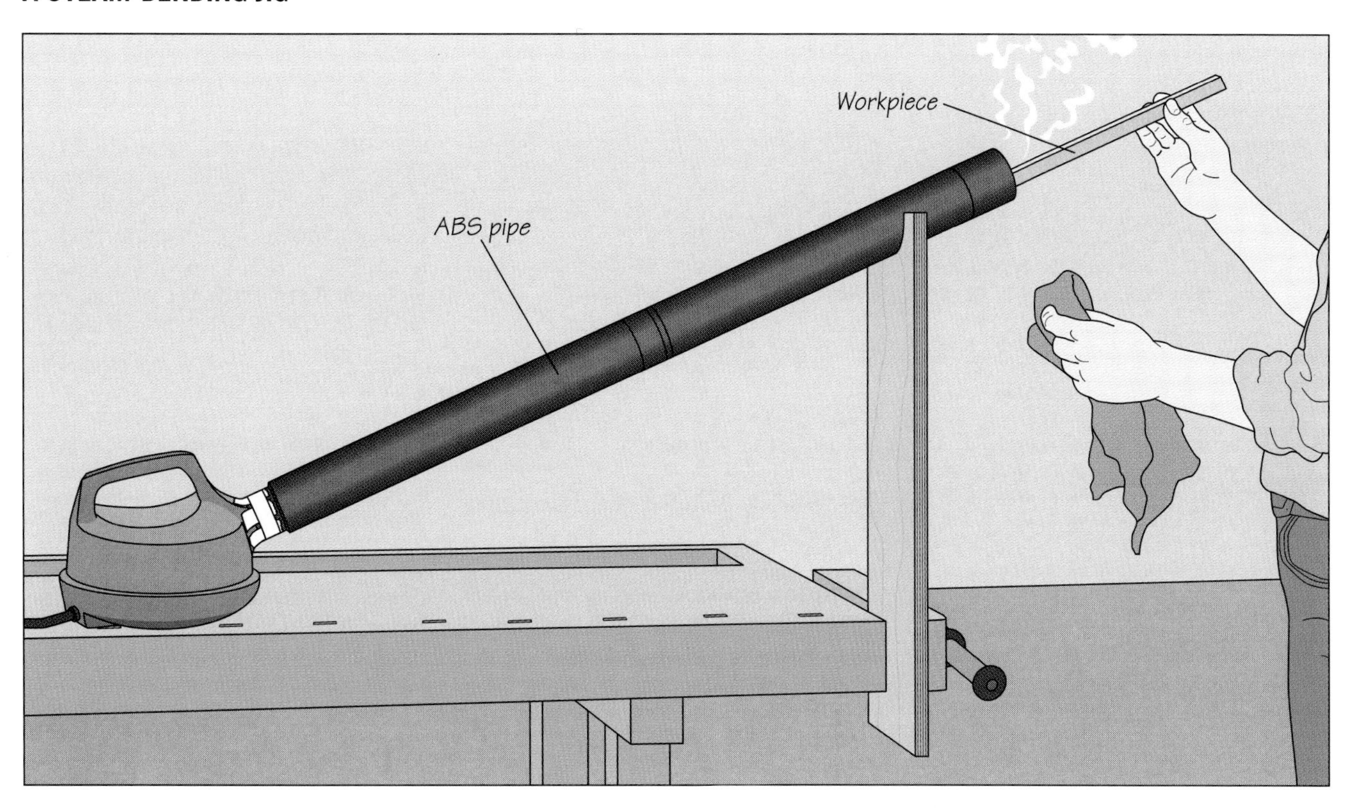

SHOP TIP

A honing jig

The simple jig shown here will enable you to sharpen plane blades without the aid of a commercial honing guide. Slip a 4-inch-long, 3/8-inch-diameter carriage bolt through the blade's slot. Fasten washers and wing nuts on both sides of the blade. With the cutting edge on the sharpening stone and the head of the bolt on your work surface, use a protractor and a sliding bevel to adjust the wing nuts so that the blade can be slid along the stone at the proper angle, typically 30°.

Steam-bending a workpiece

To steam-bend small workpieces, such as strips of molding, build the portable steamer shown above. Cut a length of ABS pipe longer than the wood you wish to bend and about the same diameter as the spout of an electric kitchen kettle; use a kettle with a round spout. Connect one end of the pipe to the spout, making a tight seal with duct tape. Support the other end of the pipe at a slight angle on a notched piece of plywood clamped in a bench vise. To use the jig, bring the water to a boil, insert the workpiece to be steamed *(above)*, and stuff a rag into the open end of the pipe to contain the steam. Let the workpiece "cook" until the wood softens; as a rough guide, allow 1 hour per inch of thickness. Refill the kettle as necessary, plugging the end of the tube temporarily to contain the steam.

GLOSSARY

A-B-C

Auxiliary fence: A wooden attachment to the rip fence of a table saw or other machine, for the purpose of avoiding accidental damage to the metal fence.

Bench dog: A round or square peg made of metal or wood that fits into a hole in a workbench to hold a workpiece in place.

Bench stop: A jig fastened to a work surface to steady a workpiece.

Bevel cut: A cut made at an angle from face to face through the thickness of a workpiece.

Biscuit joiner: A portable power tool used for cutting slots for wooden wafers in biscuit joinery. Also known as a plate joiner.

Biscuit joint: *See plate joint.*

Box joint: A corner joint featuring straight, interlocking fingers.

Carcase: The box-like body of a piece of furniture.

Chamfer: A bevel cut along the edge of a workpiece.

Chuck: Adjustable jaws of a drill that hold bits and other accessories.

Clamping capacity: The widest span of a clamp's jaws.

Clearance hole: A hole bored in a workpiece to accommodate the shank of a screw.

Collet: The sleeve of a router that holds the bit shank.

Countersink: To drill a hole that allows a screw head to lie flush with or slightly below the surface.

Cove: A concave decorative profile cut in wood, usually along an edge.

Crosscut: A cut made across the grain of a workpiece.

D-E-F-G-H-I

Dado: A rectangular channel cut across the grain of a workpiece.

Dowel: Wood pin used to reinforce certain types of joints.

Edge-gluing: Bonding boards together edge-to-edge to form a panel.

Extension table: An accessory work surface used to expand the working area of a stationary machine.

Face-gluing: Similar to edge-gluing, except that boards are bonded face-to-face.

Faceplate turning: Turning circular or cylindrical workpieces such as bowls on the faceplate of a lathe.

Featherboard: A board with thin fingers or "feathers" along one end, clamped to the fence or table of a stationary tool to hold a workpiece securely.

Fence: An adjustable guide to keep the edge of a workpiece a set distance from the cutting edge of a tool.

Fixture: Shop-made device attached to a stationary or portable power tool to increase its capacity or efficiency.

Frame-and-panel joinery: A method of assembling boards into rectangular frames; grooves along the inside edges of the frame enclose a panel.

Furring strip: A thin board that is nailed to a wall or ceiling to provide a flat or level surface; normally used for securing drywall or paneling.

Groove: A cut along the grain of a workpiece, forming a channel; frequently decorative, but sometimes part of a joint.

Headstock: The spindle attached to the motor of a lathe; holds work for spindle-turning in conjunction with the tail stock, or used alone for turning with a faceplate. See *tailstock.*

J-K-L-M-N-O-P-Q

Jig: Device for guiding a tool or holding a workpiece in position.

Jointing: Cutting thin shavings from the surface of a workpiece until it is flat and perpendicular to the adjoining surface.

Joist: A horizontal support for a floor; analogous to the rafters of a roof.

Kerf: The cut made by a saw blade.

Kickback: The tendency of a workpiece to be thrown back toward the operator of a machine or tool.

Lap joint: A joint in which matching dadoes or rabbets overlap to connect two boards.

Miter cut: A cut that angles across the face of a workpiece.

Miter gauge: A device that slides in a slot on the saw table, providing support for the stock as it moves past the blade for crosscuts; its angle can be adjusted for miter cuts.

Mortise: A rectangular, round, or oval hole cut to accommodate a tenon.

Mortise-and-tenon joint: A joinery technique in which a projecting tenon on one board fits into a cavity—the mortise—in another.

Pilot bearing: A cylindrical metal collar above or below the cutting edge that guides a router bit along the workpiece or a template.

Pilot hole: A hole bored into a workpiece to accommodate a nail shaft or the threaded part of a screw; usually slightly smaller than the shaft or threaded section of the screw. The hole guides the fastener and prevents splitting.

Plate joint: A method of joining wood in which oval wafers of compressed wood fit into slots cut in mating boards.

Pocket hole: An angled clearance hole that allows a screw head to be recessed below the surface; often used when joining rails to a tabletop.

Push block or stick: A device used to feed a workpiece into the blade, cutter, or bit of a tool to protect the operator's fingers.

Quill stroke: The length of travel of the quill of a drill press.

R-S

Rabbet: A step-like cut in the edge or end of a board; usually forms part of a joint.

Radius: The distance from the center of a circle to its outside edge; one-half the diameter.

Rail: The horizontal member of a frame-and-panel assembly. See *stile.*

Raised panel: A piece of wood that forms the center of a frame-and-panel assembly. Beveling the edges of the panel creates the illusion that the central portion is raised.

Release cut: A preliminary incision from the edge of a workpiece to a line about to be cut; enables a band saw or saber saw to cut tighter curves by facilitating the removal of waste wood.

Rip cut: A cut that follows the grain of a workpiece—usually made along its length.

Shooting board: A jig for holding the end grain of a workpiece square to the sole of a plane.

Sizing board: A jig used to cut workpieces to length, typically with a handsaw.

Snipe: A concave cut created by a jointer or planer at the end of a workpiece, the result of improper pressure or table height.

Spindle: The threaded arbor on a lathe that holds the headstock or faceplate; also the threaded arbor on a drill press or shaper that holds cutters, bits, and other accessories.

Spindle turning: Turning cylindrical workpieces held between the headstock and tailstock of a lathe.

Square: Adjoining surfaces that meet at an angle of 90°.

Stile: The vertical member of a frame-and-panel assembly. See *rail.*

T-U-V-W-X-Y-Z

Tailstock: The adjustable spindle on a lathe; used in conjunction with the headstock to hold work for spindle turning. See *headstock.*

Taper cut: An angled cut along the length of a workpiece that reduces its width or thickness at one end.

Tearout: The ragged edges produced when a blade or cutter tears the wood fibers rather than cutting them cleanly.

Template: A pattern, typically made of plywood or hardboard, used with a power tool to produce multiple copies of an original.

Tenon: A protrusion from the end of a board; cut to fit into a mortise.

Wall stud: A vertical framing member forming walls and supporting the framework of a building.

INDEX

Page references in *italics* indicate an illustration of subject matter.

ACKNOWLEDGMENTS

The editors wish to thank the following:

ROUTING AND SHAPING JIGS

Adjustable Clamp Co., Chicago, IL; American Tool Cos., Lincoln, NE; Black & Decker/Elu Power Tools, Towson, MD; Delta International Machinery/Porter-Cable, Guelph, Ont.; De-Sta-Co, Troy, MI/Wainbee Ltd., Montreal, Que.; Freud Westmore Tools, Ltd., Mississauga, Ont.; Lee Valley Tools Ltd., Ottawa, Ont.; Sears, Roebuck and Co., Chicago, IL; Shopsmith, Inc., Montreal, Que.; Patrick Spielman (jig on pp. 34-35), Fish Creek, WI; Vermont American Corp., Lincolnton, NC and Louisville, KY

CUTTING JIGS

Adjustable Clamp Co., Chicago, IL; Delta International Machinery/Porter-Cable, Guelph, Ont.; De-Sta-Co, Troy, MI/Wainbee Ltd., Montreal, Que.; Fisher Hill Products, Inc., Fitzwilliam, NH; Freud Westmore Tools, Ltd., Mississauga, Ont.; Hitachi Power Tools U.S.A. Ltd., Norcross, GA; Frank Klausz (jig on pp. 58-59), Frank's Cabinet Shop, Inc., Pluckemin, NJ; Sandvik Saws and Tools Co., Scranton, PA; Dave Sawyer and Joan Sawyer (jig on p. 46), South Woodbury, VT and Rancho Santa Fe, CA; Sears, Roebuck and Co., Chicago, IL; Shopsmith, Inc., Montreal, Que.; Skil Power Tools Canada, Markham, Ont.; Vermont American Corp., Lincolnton, NC and Louisville, KY

DRILLING JIGS

Adjustable Clamp Co., Chicago, IL; Delta International Machinery/Porter-Cable, Guelph, Ont.; Sears, Roebuck and Co., Chicago, IL; Vermont American Corp., Lincolnton, NC and Louisville, KY

TURNING JIGS

Adjustable Clamp Co., Chicago, IL; Delta International Machinery/Porter-Cable, Guelph, Ont.; Lee Valley Tools Ltd., Ottawa, Ont.; Vermont American Corp., Lincolnton, NC and Louisville, KY

GLUING AND CLAMPING JIGS

Bill Bivona (jig on p. 85), Hardwood Design, Inc., Slocum, RI; Martha Collins (jig on p. 84), Lost Mountain Editions, Ltd., Sequim, WA; Lee Valley Tools Ltd., Ottawa, Ont.

SANDING JIGS

American Tool Cos., Lincoln, NE; Black & Decker/Elu Power Tools, Towson, MD; Delta International Machinery/Porter-Cable, Guelph, Ont.; Vermont American Corp., Lincolnton, NC and Louisville, KY

TOOL EXTENSIONS AND TABLES

Adjustable Clamp Co., Chicago, IL; American Tool Cos., Lincoln, NE; Delta International Machinery/Porter-Cable, Guelph, Ont.; Hitachi Power Tools U.S.A. Ltd., Norcross, GA; Lee Valley Tools Ltd., Ottawa, Ont.; Steiner-Lamello A.G. Switzerland/Colonial Saw Co., Kingston, MA; Vermont American Corp., Lincolnton, NC and Louisville, KY

STORAGE DEVICES

Adjustable Clamp Co., Chicago, IL; American Tool Cos., Lincoln, NE; Great Neck Saw Mfrs. Inc. (Buck Bros. Division), Millbury, MA; Leonard Lee (rack on p. 121), Lee Valley Tools Ltd., Ottawa, Ont.; Sandvik Saws and Tools Co., Scranton, PA; Vermont American Corp., Lincolnton, NC and Louisville, KY

SHOP AIDS

Adjustable Clamp Co., Chicago, IL; Campbell Hausfeld, Harrison, OH; Delta International Machinery/Porter-Cable, Guelph, Ont.; Hitachi Power Tools U.S.A. Ltd., Norcross, GA; Taylor Design Group, Inc., Dallas, TX; Vermont American Corp., Lincolnton, NC and Louisville, KY

The following persons also assisted in the preparation of this book:

Lorraine Doré, Réjean Garand (Atelier d'Ebénisterie Réjean Garand Enr., St-Rémi, Que.), Graphor Consultation, Claude Martel, Geneviève Monette, Alain Morcel (Les Réalisations Loeven-Morcel, Montreal, Que.)

PICTURE CREDITS

Cover Robert Chartier
6, 7 Robert Holmes
8, 9 Michael Morissette
10, 11 Ron Levine